MONOCHROM'S ARSE ELEKTRONIKA ANTHOLOGY

Editors: Johannes Grenzfurthner, Günther Friesinger, Daniel Fabry, Thomas Ballhausen
Publisher: RE/SEARCH
Copy editing: Melinda Richka
Layout: Daniel Fabry, Anika Kronberger
Arse Elektronika logo: Tokyo Farm
RE/SEARCH Staff: V. Vale, Marian Wallace, Seth Robson, Robert Collison, Jared Power, Alex Lavine, Cayla Lewis, Stellar Kutchins, Michael Raines, Joanna Sokolowski, Ilana Fried, Joe Donahoe

This publication was supported by the Department of Art Funding / City of Vienna, Austria.
Printed in California, USA.

LETTERS, ORDERS & CATALOG REQUESTS TO:
RE/SEARCH PUBLICATIONS
20 ROMOLO #B
SAN FRANCISCO, CA 94133, USA
PH (415) 362-1465
email: info@researchpubs.com
www.researchpubs.com

Arse Elektronika 2008 was organized by monochrom.
www.monochrom.at/english

Curators: Johannes Grenzfurthner, Günther Friesinger, Thomas Ballhausen
Financial supervisor: Günther Friesinger
Corrections of all kinds: Evelyn Fürlinger
Web supervisor: Franz Ablinger
Press: Roland Gratzer
Web design: Anika Kronberger
Technical supervisor: Daniel Fabry
Co-organizers: David Fine, Carol Queen

Many thanks to the supporters of Arse Elektronika 2008: Laughing Squid, Center for Sex and Culture, Chaos Computer Club, Simone Davalos, David Calkins, Melinda Richka, Mediapathic Steen and the Department of Art Funding, City of Vienna, Austria.

CONTENTS

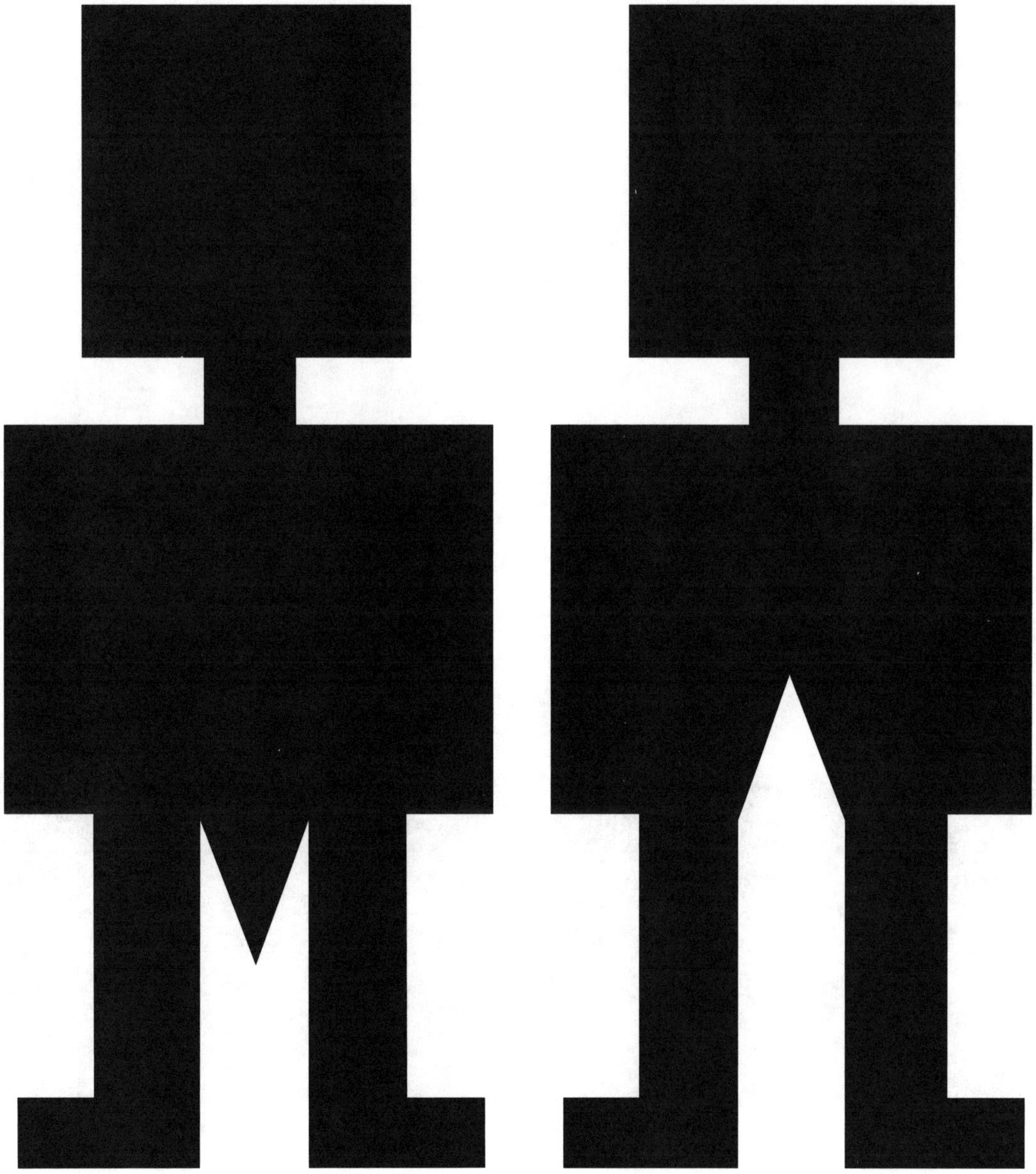

monochrom's Arse Elektronika 2008

DO ANDROIDS SLEEP WITH ELECTRIC SHEEP?

CRITICAL PERSPECTIVES ON SEXUALITY AND PORNOGRAPHY IN SCIENCE AND SOCIAL FICTION

Picking up where we left off with the first part of our successful series, this anthology falls under the motto 'future' - and the ways in which the present sees itself reflected in it. Maintaining a broadened perspective on technical development and technology while also putting special emphasis on its social implementation, this year's conference focuses on Science and Social Fiction.

The genre of the 'fantastic' is especially well suited to the investigation of the touchy area of sexuality and pornography: actual and assumed developments are frequently depicted positively and approvingly, but just as often with dystopian admonishment. Here the classic, and continuingly valid, themes of modernism represent a clear link between the two aspects: questions of science, research and technologization are of interest, as is the complex surrounding urbanism, artificiality and control (or the loss of control). Depictions of the future, irregardless of the form they take, always address the present as well. Imaginations of the fantastic and the nightmarish give rise to a thematic overlapping of the exotic, the alienating and, of course, the pornographic/sexual as well.

Johannes Grenzfurthner (Head of conference), Günther Friesinger, Daniel Fabry, Thomas Ballhausen

monochrom
Art-tech-philosophy-collective

Rudy Rucker

SEX AND SCIENCE FICTION

Science fiction is a mountain of metaphors, a funhouse of crooked mirrors that give us new views of our actual world.

From our genes' point of view, we're meat-based landcrawlers to ride around in. Imagine little double helices lounging in the hammocks of your cells. What makes us especially useful is that, now and then, we spawn off new landcrawlers with copies of the passenger genes, carrying them ever forward through time.

Putting the same point differently, if living organisms weren't obsessed with sex none of us would be here. We're each a link in a chain of generations, we're dangling dollies on a slimy macramé of a trillion umbilical cords.

Of course we enjoy sex for more immediate reasons than reproduction: erotic pleasure, the orgasm, and partnership bonding. The last one is important. That's why we talk about *making love*. We're wired so that love readily grows from the sex act.

Certainly, if reproduction were the only reason for sex, you wouldn't be having so many orgasms. How many? Math time! Suppose you live to your eighties, and that you have seventy years of sexual activity, which makes for about 3,500 weeks. If you're energetic enough to average three pops a week for seventy years, you're talking about something on the order of ten thousand orgasms. All that brain-flashing to bring forth at most a couple of kids! 'Oooo Mommy, you mean you and Daddy did that *twice*?'

So how about science fiction and sex? Where have we been, where are we headed, and how much further can we go?

One sex story I always think of is Samuel Delany's, 'Aye and Gomorrah,' about a cadre of spacers who've been surgically altered so that their crotches are as featureless as those of a plastic Barbie doll's. Why? Given the amount of mutating radiation that these astronauts absorb in their space-stations, it would be too dangerous to allow them to reproduce. In the story, there are people who are sexually obsessed with the Barbie-smooth spacers. These fetishists are called *frelks* - a great word.

In this context, I also think of a particular story about people being sexually attracted to aliens, 'And I Awoke and Found Me Here on the Cold Hill's Side,' written by Alice Sheldon, under her nom de plume James Tiptree, Jr. Upon seeing aliens, the story's characters have a surprising and overwhelming sense of lust. Kind of like how some of us may react to our first sight of a gay pride parade! Ah, those six-foot-tall honking-loud brides...

One reason we're attracted to sex with other people is simply because they're different. Gender isn't necessarily an issue. That's the core idea in both the Delany and the Sheldon stories: otherness is a turn-on. And any other person is, for all practical purposes, an alien, if you really think about it.

Note that it's not just the *difference* that turns us on, it's the idea that there's an *intelligent mind* inside the different body. Another mind that mirrors you, a mind you can in fact pair up with for an endless regress of mutual reflections.

There's a major difference between sex with a person and sex via media. In sex with a person, you're talking about emotion, the positions of your limbs, touch across large skin areas - about tastes, scents and pheromones. A candle by the bed is nice, but you can just as easily make love in the dark.

In media-based sex, we're reduced to visual images, perhaps enhanced by recorded sounds. But there's no emotion, touch, tastes, or smells. And text-based sex is even more abstract.

I'm a little sorry to see the decline of text-based pornography. It used to be in every corner store, and now you hardly see it - although it can be found online. In the 1970s, I had a bar-fly friend who was paid by the hour to write porno novels in an office in downtown Rochester, New York. I thought he was cool. A real writer!

Still on the theme of sex with aliens, my novel *The Sex Sphere* features a giant ass from the fourth dimension. She's called Babs. She has eyes, breasts, a mouth, a vagina - but no limbs. She can fly, she's into nuclear terrorism, and her ultimate goal is to utterly destroy our universe. Have any of you ever dated her? The book was recently reissued by E-Reads.

One of the earliest bizarre SF sex stories that I read was in Philip Jose Farmer's 1950s anthology, *Strange Relations*. I'm thinking of his story, 'Mother' in which a stranded space-explorer finds shelter within a cavity in a meaty plant. The plant - or perhaps it's an animal - feeds him food and bourbon, nursing him along. And it turns out that the astronaut is expected to attack a certain area of the plant's womb, which will catalyze her into a pregnancy, enabling her to bear young. And after his attack the mother-plant will eat him. In a way, it's an incest story, but looked at differently; it's also a story about retreating into a cocoon.

Think of a person alone with their computer - whether they're viewing internet porn, having sex-talks in chat-rooms, or playing erotic roles in a multiple-user videogames. Or think of people lying in Matrix-style jelly-pods with their brains plugged into a group virtual reality.

I find these scenarios sad. In 'Mother,' the character at least has the ability to fecundate the surrounding blob - but what can you as an individual do to the internet? What can you do to some vast virtual reality that you're duped into spending all your time with?

Well, in the case of the internet, at least you can post comments, upload videos, start a photostream, run a blog. And maybe, if you're lucky, you can galvanize another human into meeting you face to face.

It's always important to remember that computers are dead and boring compared to our fellow humans. Even if there's a human on the other side of the computer output that you're interacting with, the machine is still between you, even more isolating than - you should pardon the expression - a glory-holed toilet-stall wall.

For a little while, people were talking about having sex via the internet by means of computer-operated sex toys. It's doable, but who wants to bother? It's the skin that matters, the breath, the eyes, the voice.

As a partial improvement, in my novel, *Freeware*, I had sex toys that were made of flexible and intelligent plastic that could move on its own. I called the material 'piezoplastic,' and it had become rather intelligent due to a wetware mold infestation. Bigger chunks of the fungus-dosed piezoplastic were autonomous and vicious beings called moldies - and those who loved them were known as cheeseballs. Moldies would take control of a cheeseball by inserting a small slug of their plastic into the human's skull, and the sluggie would run the person like a robot remote. You might call this an objective-correlative for sexual obsession.

As an SF writer, I wonder if there could be a non-plastic and purely biological medium for enjoyable remote sex. Certainly a sex-toy would be more congenial if it were made of a human tissue culture instead of plastic. Ideally the seed cells for the tissues would come from your lover's body, so that the smells and pheromones are just right. Actually, Bruce Sterling and I wrote a story called 'Junk DNA' in which these little jobbies were called Pumptis.

Of course, for full satisfaction, the personal-intelligence touch is needed. You want a way to project your mind into that remote Pumpti that your darling is going to use - and vice-versa. Well, we can do that via quantum entanglement, no prob. Everything's easy in science fiction.

While your partner is getting it on with your Pumpti, you'll be diddling the Pumpti that he or she gave you. And, even better, you'll be projecting your consciousness into the remote Pumpti and into your partner's mind as well.

Great. But, wait - this doesn't sound all that different from phone sex. Or love-letters.

Basically, remote sex is boring. There's no substitute for face-to-face. Let me say a little about possible SFictional amplifications for in-person encounters. For instance in my *Ware* novels, there's this drug called merge. Lovers get into a bathtub called a love puddle, they splash on the merge, and their bodies melt and flow together - making a happy glob of flesh with four eyes on top. After an hour or so, the merge wears off, and the couples' body shapes return.

I like to think of telepathy as a sexual enhancer. I already mentioned that it's exciting to have your own mind mirrored in someone else's, even as you're mirroring then and so on forever. Suppose that the mirroring is through a direct brain contact. It's easy to suppose that the feedback could flip into a chaotic mode, generating fractal strange attractors. It would take a bit of delicate maneuvering to avoid spiraling into the fixed-point attractor of a brain seizure.

Here's a longer passage about this, drawn from my novel *Saucer Wisdom* (I also discuss this idea in my novel *Hylozoic*).

> At first it's mellow. Larky and Lucy lie there side by side on the floor, smiling up at the ceiling, thinking colors and simple shapes. Blue sky, yellow circle, red triangle. Now Larky puts his hand in front of his face, stares at it, and the image goes over to Lucy. But Lucy isn't able to see the hand yet. She can't assimilate the signal. 'You try and send a picture to me,' says Larky. He doesn't say the words out loud, instead he *imagines* saying them - he subvocalizes them as it were - and Lucy is able to hear them. Words are easier than pictures. Lucy stares at her piezoplastic bracelet, fixating on it, sending the image out. Larky can't get it at first, but then after a minute's effort, he can. Eureka!
>
> 'You have to let your eyes like sag out of focus and then turn them inside out, only without physically turning them, you wave?' explains Larky none too clearly. 'It's sort of like the trick you do in order to see your eyes' floaters against the sky. You're looking far away, but you're looking inside your head.'
>
> So now Larky and Lucy can see through each other's eyes, but then Larky glances over at Lucy and she looks at him and they get into a feedback loop of mutually regressing awareness that becomes increasingly unpleasant. It's kind of like the way if you stare at someone and they stare back at you, then you can read what they think of you in their face, and they can read your reaction to that, and you can read their reaction to your reaction, and so on. It gets more and more intense and pretty soon you can't stand it and you look away.

But with a direct brainwave hookup, the feedback is way stronger. In fact it's like what happens when your point a video camera at a TV monitoring what the camera sees. Lucy's view of Larky's face forms in Larky's mind, gets overlaid with Larky's view of Lucy and bounced back to Lucy, and then it bounces back to Larky, bounce bounce bounce back and forth twisting into ragged squeals.

Lucy and Larky are starting to tremble right on the point of going into some kind of savage epilepsy-like fit - but Larky does a head-trick that makes it stop.

Larky's method for stopping the feedback is like one of the things you can do with the video camera to keep the TV screen from getting all white: you zoom in on a detail. You find a fractal feather and amplify just that. In the same way, Larky shifts his attention to a little tiny part of his smeared-out mouth, a little nick at the corner, and a soon as that starts to amp up, he shifts over to a piece of Lucy's cheek, just keeps skating and staying ahead of the avalanche. Lucy gets the hang of it too, and now they're darting around their shared visual space.

Larky and Lucy slowly develop a language for transmitted emotions. Part of the trick is to keep a low affect, to speak softly as it were. If you scream a feeling, it bounces back at you and starts a feedback loop. You can think a scream, but you have to do it in a calm low-key way. The way Lucy puts it, 'Just go 'I'm all boo-hoo,' instead of actually slobber-sobbing.' So, pretty soon Larky and Lucy are good at sending emotions in a gentle chilled-out kind of way.

Not everyone can remember to stay chilled out and to not stare into the feedback. The other big hurdle is to make the signals readily comprehensible. Larky and Lucy were able to communicate quite easily because they knew each other really well: they're lovers and best friends. But what happens when you try and link with a relative stranger? None of his/her references and associations make sense.

The trick turns out to be to first exchange copies of your lifebox contexts. As well as using the analog signals of the superquantum brain sensors, you also use standard hyperlinks into the other user's context. The combination of the two channels gives the effect of telepathy.

Some couples become addicted to the dangerous intensity of skirting around the white hole of feedback; of bopping around right on the fractal edges of over-amplification, on the verge of tobogganing towards the point-attractor of a cerebral seizure. Fortunately you could always shut off your telepathy. With practice, Larky and Lucy had learned to skate around the singular zones, enjoying the bright, ragged layers of feedback.

Coming back to the concept of sex as reproduction - what if you were engendering something more than a child? In a couple of my novels, I've had couples who somehow save our universe as a side-effect of their love-making. Father Sky and Mother Earth is an old legend that expresses something fundamental: sex as creation.

Here's a version of this from *Spacetime Donuts*.

He was floating, a pattern of possibilities in an endless sea of particulars.

'Be the sea and see me be,' the words formed … somewhere.

He let his shape loosen and drift to touch every part of the sea around him, a peaceful ocean like a bay at slack-tide on a moonless summer night…peaceful, while in the depths desperate lives played out in all the ways there are. Taken all together, the lives added up to a messageless phosphorescence, a white glow of every frequency.

'And are you here?'

'As long as you are.'

'Can we go further?'

NOW

And there's a scene like this in *Postsingular*.

They undressed and began making love. They had all the time in the world. Everything was going to be all right. At least that's what Jayjay kept telling himself. And somehow he believed it. He and Thuy were one flesh, all their thoughts upon their skins. Their bodies made a sweet suck and push. The answer was before them like a triangular window.

Jayjay had been too tense and rushed to teep the harp before. But now - now he could feel the harp's mind. She was a higher order of being, incalculably old and strange. She knew the Lost Chord. She was ready to teach it to him. Jayjay and Thuy melted into their climax, they kissed and cuddled. Jayjay got up naked and fingered the harp's strings. They didn't hurt his fingers one bit.

The soft notes layered upon each other like sheets of water on a beach with breaking waves. Guided by the harp, Jayjay plinked in a few additions, thus and so. And, yes, there it was, the Lost Chord. Space twitched like a sprouting seed.

And with that, the harp was gone.

No matter. The sound of the Lost Chord continued unabated, building on itself like a chain reaction, vibrating the space around them. Jayjay smiled at Thuy. He had a sense of endlessly opening vistas.

'You did it,' said Thuy. 'You're wonderful.' She wasn't talking out loud. Her warm voice was inside his head. True telepathy. Jayjay had unrolled the eighth dimension. He and Thuy had saved the world.

Sex is everything.

Isaac Leung

THE CULTURAL PRODUCTION OF SEX MACHINES AND THE CONTEMPORARY TECHNOSEXUAL PRACTICES

From the simple electronic vibrator to the complexities of cybersex, sex and technology have always been intersected. The dynamic relations between sexuality and technology are constantly changing along with the ways in which human beings achieve psychological and bodily pleasure through these devices. In this paper, I will examine the production of sex machines as a cultural artefact and evaluate how sex machines are being produced and culturally imagined. In order to find out how sex machines have culturally redefined sexuality, technology, gender and the body, I will document different values, beliefs, ideologies and practices engaged during the invention and production processes, especially those of fucking-machines, teledildonics and sex robots.

Types of sex machines

This paper focuses on three kinds of sex machines that have been produced in recent years: fucking-machines, teledildonics and humanoid sex machines. Fucking-machines, which are intended for performing penetrative sex, are electrically operated thrusting and spinning devices with phallic attachments that imitate or respond to body movement. Teledildonics, a term first used by sociologist Theodor Holm Nelson, was conceptualized as an integration of sex and telepresence that essentially refers to remote-controlled sex. In recent years, teledildonics usually denote a sex device that is controlled by computer networking systems. The sex robot is an artificially created agent that mechanically resembles a human and is made specifically to assist or replicate real humans in the performance of sex. Research on humanoid sex machines is increasingly popular in the sex industry, while fictional sex robots are widely imagined in films and art.

Identification of the 'sexual field'

Based on the Bourdieusian concept of field and Goffmanian analysis of social psychology, 'sexual field', a term created by Dr. Adam Isaiah Green, examines how individual agents develop a reflexive relationship to their sexual practices with the possession of 'erotic capitals'. 'Sexual field' enables us to study the power relations of different erotic agents and characterize collective sexual life (Green, pp.25-50, 2008). The sexual field of sex machines in this paper includes the independent production of fucking-machines, the production of fucking-machines in pornography and sex toy industries, the artistic imagination and industrial production of sex robots, and the cultural production of sex machines within major institutions that are dedicated to the study and research on sex machines. Producers in the field of sex machines were interviewed in order to find out how sex machines were being technically produced and culturally situated.

Independent productions of sex machines

The phenomenal achievements in mechanical flexibility and electrical and tele-networking technologies have proliferated among the independent inventors of fucking-machines and the open-sourcing of sexual programming. Mechanical and electrical knowledge have not been limited to the professional but have become accessible to the general public in DIY culture today. The independent productions of fucking-machines are situated at the fringe of the sex toy industries. The convergence of garage laboratories and sexual innovation, where tools, hardware, electrical appliances and sex toys are readily available, has created an outlet for many independent creations. Some of the most extensive research on independent fucking-machines is documented in Timothy Archibald's book, *Sex Machines: Photographs and Interviews*. Archibald traveled to rural towns and suburbs across the United States between 2003 and 2005. He discovered more than thirty do-it-yourself fucking-machine enthusiasts through his research. His book includes a series of documentary photographs and interviews that are aimed not only at artistic expression, but also sociological investigation (Archibald, pp104, 2005).

Apart from the mechanical innovations, advanced computer-mediated communicating (CMC) technology allows Internet users to interface with social networks for exchanging sexual information and sexual sensation. Kyle Machulis, the first dildo maker in Second Life and a full-time robotician, is known for his innovations in teledildonics and video gaming programming which are being applied for sexual purposes. He is on the leadership council of the International Game Developers Association and specializes in sexual applications for video games. His projects include *Slashdong*, a blog about the electronic and mechanical engineering of sex toys, *opendildonics.org*, an open source teledildonics wiki; and *MMOrgy*, a website that advocates sexual activities that are being applied in the MMOG (Massively Multiplayer Online Gaming) community.

Industrial productions of sex machines

Developments in new technologies have provided the sex industry new ways to produce, market and deliver their sexual products. From the invention of analog devices such as phone, film, and photography to digital inventions such as high-speed Internet connections and mass data storage systems, the sex industry has always been closely intertwined with new technologies. Sex machines, a type of sex object or device that is inseparable from technology, has been widely utilized by entrepreneurs in the porn and sex toy industry. For example, the growing number of independently invented fucking-machines and the booming of internet pornography have inspired a former PhD student of Columbia University in finance, Peter Acworth, to open *Fucking Machines* in 2000. *Fucking Machines* is the first porn site that is entirely dedicated to human interactions with thrusting machines. It is one of the projects under the umbrella of *Kink.com*, a major porn company in the United States with a production studio that employs over a hundred people located at the former San Francisco Armory. Likewise, the computer-networked sex machines have also been widely produced in the sex industry. Alan Stein, the owner of Thrill Hammer and the co-founder of *Sex Machine Cams*, is the pioneer of commercial teledildonics inventions. Based in Seattle, Thrill Hammer offers custom-made teledildonics services and online retailing of fucking-machines that are produced by different manufacturers, while *Sex Machine Cams* is the first porn company which specializes in interface designs so that users can control fucking-machines in real time via the internet.

Apart from fucking-machines and teledildonics, advanced humanoid sex machines that resemble the human body structure through mechanical, electrical and (or) artificial intelligent agents are widely conceptualized and produced by sex industrialists and artists. I interviewed, Michael Harriman of *First Androids* about his sophisticated humanoid sex dolls that have the capability to generate human body temperatures and perform bodily and respiratory movements. Located in Nuremberg, Germany, *First Androids* is the only company that provides online orders for custom-made humanoid sex dolls. Every product comes with unique body features and functions. Apart from the industrial creations of humanoid sex machines, artists such as Shulea Cheang actualize robotic imaginations into video art. Shulea Cheang created bioengineered humanoid robots that are also known as 'I.K.U. Coder'. Cheang's *I.K.U.* is a pornographic art film that portrays new forms of sex that have been invented by a futuristic corporation. Besides 'F Coder', Shulea Cheang also envisioned a future orgasm decoding technology that allows consumers to download and experience orgasms without bodily contact through an 'I.K.U. Chip'.

Institutional productions of sex machines

On top of the independent, industrial and artistic production of sex machines, the 'sexual field' of sex machines is also mediated by the institutional factors that motivate the cultural production of sex machines. Non-governmental institutions in contemporary societies have recently been concerned with proclaiming, promoting and legitimizing certain ideologies, arrangements and practices. Knowledge and people's conceptions are socially constructed through institutional processes.

(Berger & Luckmann, 1976) Just like any cultural product, sex machines are partially mediated by different institutional agents such as academic and curatorial practices. In this paper, *Arse Elektronika*, a sex-oriented conference, will be examined because of the projects it has done on sex machines. In recent years, many academic conferences have focused on different aspects of sex and technology, including psychology, public health, sex culture and education. *Arse Elektronika*, which is sponsored by the Department of Art Funding in Austria, is a sex and technology conference that in 2007 aimed at exploring pornography and sex machines. Organized by an 'art-tech-philosophy' collective, *monochrom*, *Arse Elektronika* was held in San Francisco, and located next to the high tech industries of Silicon Valley, which has a long history of sexual diversity and technological advancement. Among many scholars who concern about sex and technology, there was David Levy, an artificial intelligence researcher, whose PhD thesis is about the human-robot relationships in love and sex. He is currently one of the owners of a computer game company in London. He is also the president of the International Computer Games Association. His prediction concerning robotic marriage by the year 2050 in *Love + Sex with Robots* has captured major media attention, including The New York Times, CNN and NBC.

Different agents in the 'sexual field' mentioned above mediate the cultural productions of sex machines where they constantly define and re-define the culture of sex and technology. In-depth interviews were conducted in order to find out how different agents have conceptualized, identified and operated the productions of their sex machines.

Techno-fetishism of sex machines

One of the main patterns that have been seen in the process of the productions of sex machines is 'techno-fetishism'. In order to understand how this pattern is being articulated in the productions of sex machines, it's important to understand the cultural meanings of technology and fetishism. Essentially, the meaning of technology is the application of different scientific data to achieve different practical ends. The 'technology' that is being examined here goes beyond its material nature. French sociologist, Marcel Mauss has positioned 'technology' into the sociological domain; he sees all 'objects' of technology as products of a 'total' social relation and the 'invention' of technologies represents not only the ability to 'solve a mechanical problem' but 'the processes of imagination' in society (Mauss, pp34, 2006). The 'totality' of 'technology' that is being fused with 'fetishism' in the productions of sex machines represents not only the instrumental value of the artefact itself, but also the sociological and cultural relations that the machines engendered. Different types of sex machines in the 'sexual field' entail different techniques and technologies; correspondingly, each type of sex machines is situated in a specific social and cultural condition that in return shapes the design and technological process.

What are the effects of 'technology' when it is 'fetishized'? 'Fetishism' in Marx's definition is that objects (commodities) are exchanged while the 'use-value' and 'labour value' are being effaced by capitalism, thus objects are being seen to have power over labor. Jean Baudrillard further elaborated 'fetishism' by using a semiological approach. Besides the exchange value, 'sign-value' is also generated through the 'display of commodity' (Kellner, pp21, 1994), Baudrillard concluded that consumers construct their own identity and lifestyle by 'fetishizing' the signified of objects. Though R. L. Rutsky in *High Techne* made a sound connection between the two, the fetishism of technology cannot be fully explained by the 'commodity fetishism' postulated by Karl Marx and Jean Baudrillard. Rutsky illustrated that the fetishism of technology in contemporary societies 'extends beyond the fetishism of particular high-tech object', which in other words is the extension of the instrumental functions and the non-instrumental aesthetics of the object self. He explained that the 'very idea of high-technology is itself fetishized' and the 'idea' of technology represent a 'mysterious life' of its own. He associated techno-fetishism with Karl Marx's perspective on commodity fetishism and Baudrillard's idea on signified 'style', in which objects in modern societies are not being seen in terms of their material value and their production and distribution factors, but with the idea of 'mysterious life' and the 'complex logic' that is being signified in the high-techness of the object self (Rutsky, pp130, 1999). In this paper 'Techno-fetishism' will be noted as one of the patterns that are found during the process of the production of sex machines. It is not only aimed at articulating the functional value of the technology that producers are attracted to, but also refers to the social and cultural meanings of the specific technological style that is being signified for different kinds of sex machines.

American garage - the suburban machines

When Henry Ford's Model A was being introduced in 1927, 'garage', a structure that is usually independent of the house was created for automobiles. Apart from sheltering the automobile, many garages in America are used for storage of tools and as workspaces for home improvement projects. Garages not only became part of the American family's automotive lifestyle, but became a location for home inventors, a significant element in industrial innovation. People like Steve Jobs of *Apple*, Bill Hewlett and Dave Packard of *Hewlett Packard* and Walt Disney of *Disney* invented their first products in their own garages. According to Thomas Roche, the Public Relation Manager of *Kink.com*, there were no sex devices that were designed to thrust ten to fifteen years ago, and his business was highly inspired by the independent-made 'fucking machines'.

> 'I think there's a very important aspect of amateur sex machines that was present in the early fucking machine sites, it was sort of an exciting way to create something new and kinky that also involves a great degree of craftsmanship. On my panel at the *Arse Elektronika*, we were talking very much about this idea that amateur inventors who put a lot of energy to make fucking machines, that requires a lot of complex interactions with the machinery, they put a lot of love and energy into it. There's something really interesting about that' (Roche, personal interview, 2008).

The rapidly rising popularity of fucking-machines in this decade originated from and was inspired by independent garage inventors. Scattered throughout small towns in cities like Champlin, Minnesota and Kansas City in the United States, fucking-machines are being invented in many suburban garages. In my interview with Timothy Archibald, he talked about his first exchanges with fucking-machines inventors. He was amazed by the fact that the inventor was an ordinary suburban man with grown kids and wife. He said,

> 'You saw how he (the inventor) had taken over this invention of his that he was so passionate about; he had taken over the garage where he made the machine. And I just thought it was fascinating to see this guy so passionate about mechanics. He's just a normal guy so I thought this would be an interesting collection of people to photograph. I could find other people like him' (Archibald, personal interview, 2008).

This passion about mechanics is almost a prerequisite for the garage fucking-machines inventors who appear in Archibald's book. For examples, one fucking-machine inventor, Dwaine Baccus, from Emmett, Idaho, believes that building and operating a machine that he built himself can be 'on the level of a sexual experience', 'an aphrodisiac of his mind' and 'all his senses'. He thought that the intense pleasure of building fucking machines had a lot to do with the fact that he built them himself. Ironically, Baccus didn't make machines for his own sex partners, but enjoyed seeing other couples use and test his machines. He explained that his creations were a combination of 'creative needs' with 'sexual components'. To build these machines was 'a way to express himself sexually'. (Archibald, pp20, 2008) Another garage inventor James Vermeer of Victorville, California thought that building fucking machines was all about 'the wonder of gears, bearings and housing'; the pleasure was to see these things all come together and work perfectly (Archibald, pp22, 2008). Similarly, Ruiin, a former airplane mechanic who built a series of Gothic fucking machines, states in Archibald's book that he was not into the orgasm when he had sex and making the sex machines was similar to this feeling. He said, 'The machine was like that for me: I really enjoyed making it, working out the details and the design, finding ways to do it affordably, thinking it through. But using it was not really anything special.' These similar responses made Timothy Archibald wonder why independent fucking-machines inventors were so passionate about the mechanics. He once asked an inventor and instead of getting a direct answer, he began to understand this mechanical inclination through the sound that was generated by the fucking-machines. He thought 'there was a feeling with the sound, the electronic buzz, and a powerful thing that was going to do whatever it did no matter what'. He described that this realization reminded him of a human being, but it was emotionally and mechanically stronger and faster and more powerful than any human being and 'it was not going to stop.' Many of the independent inventors didn't think about practical ends before they made the fucking- machines; a lot of the time they were preoccupied by sex and the fact that they could invent something. Obviously many of the garage fucking-machines inventors were gratified by the mechanical nature of the fucking-machines, the sensations generated by seeing, touching and hearing the machines and the process of tooling different mechanical components.

Apart from those who only made the fucking-machines for private consumptions, there are many who took their inventions into the market. In my interview, Timothy Archibald said,

'In America, there's a belief that you can get rich, you can invent something to get rich. There's Apple computer, there's light bulb, there's a thing in America that you can come up with something, and it is going to put you on a Jay Leno show, you will be talking to Jay Leno and you're going to be rich' (Archibald, personal interview, 2008).

One of the successful examples is Rick of Spindoll Manufacturing and Sales. He started his business by inventing three fucking machines and shooting home pornography. His fascinations with the fact that his machines can make woman 'come faster and harder', and they 'get them (women) off better than anyone else's machines' pushed him to do live demonstrations of his inventions with his wife Kristy for a local swinger club in Henderson, Nevada. His invention 'Orgasmo' made him famous in the world of sex machines. It became one of the most highly priced fucking machines in the market. Some inventors didn't attempt to make luxury sex Fmachines; they wanted to make their inventions affordable and easily available. For example, New Orleans-based Ken Cruise, who has a day job at a major retailer, works nightly in the garage for his family-run sex machines business called *Ken's Twisted Mind Inc.*. He put his invention 'Hide-a-Cock' on eBay for USD250 and was able to immediately make a sale. The week after he made 30 more transactions. Another garage inventor Scott Ehalt from Champlin, Minnesota, created 'Ultimate Ride' by using his kitchen table. He brought his invention to the Bank of America and tried to explain his business plans. The business wasn't as big as he planned, though he was still producing machines when he received his first orders online. These examples represent a common practice of entrepreneurship that is deep-seated in the American garage culture; the practical ends for the inventions were not confined to the functionality of the fucking-machine itself, they also engender the possibility of setting up businesses for innovative products that are not yet being produced in the market.

How is 'technology' being 'fetishized' in the suburban context of the independent invention of fucking-machines? Needless to say, many of the garage inventors were fascinated by the material nature of machines and mechanics; the process of articulating the 'technology' itself was symbolically equivalent to the process of having sex. It was not hard to find out through my interviews that during the process of producing the fucking-machines, a way to sexualize machinery came about.

Timothy Archibald came up with his project by studying the American suburban lifestyle. He said during his interview, 'I wasn't inventing sex machines, but I live in the suburbs and I have a kind of unsensational life, I got kids and wife and taking kids to school. So I liked the idea of suburban mundanity, like the truly dreary unsexy things that you do in you life.' The social and spatial conditions of American suburbia represent the 'familial isolation through a lack of public space and through an emphasis on home maintenance and home-centered entertainments' (J. Miller, pp393, 1995). The lack of public spaces such as café, central plaza, train stations, or movie theatres in suburbia limits social interactions with friends and strangers. Family and home oriented activities are more or less the only choices left since many suburban residents have nowhere to go in the suburban setting. Since public spaces are 'decentralized' from suburban house, automobile trips became necessary for many daily tasks (J. Miller, pp395, 1995). The garage, a private space originally designed to shelter an automobile also became the ideal space for home entertainments and maintenance. The mechanical objects in the garage range from the automobile to the lawn mower to the hobbyist's tools, things that became essential elements in suburbia's everyday needs along with concomitant social and cultural lifestyle concerns. The suburban is someone who uses, remodels and invents machines and mechanical objects not only because of the object's material functionalities, but also due to the process of usage, the remodeling and invention of those machines and mechanical objects themselves are being 'fetishized' into the suburban social and cultural life. The 'technology' that is being 'fetishished' for the independent fucking-machines inventions is the signified style of the American suburban life, i.e. the use of garage and machines and the tooling of mechanics due to the unique suburban social and cultural conditions. Besides those who are only making the machines for self-(sexual) entertainments, there are garage inventors who took their fucking-machines into the market-place. Those independent inventors brought the 'technology' into the entrepreneurial dimension. The 'technology' in those productions is the signified American dream of that turning one's ideas, visions and creativity into a mass-produced product; anyone who wants to set up a business has a chance to try. Fucking-machines are being produced in the garage, at the same time, suburbian's (sex) life is also being re-invented, used, and remodeled through the process of the production of these D.I.Y. fucking-machines, and these machines embody the broader American suburbanist sexual modernity.

Open-sourcing sex in teledildonics

Many of the garage-made fucking-machines are being discussed, promoted and sold on Internet forums and shopping sites like *eBay*. Technological advancements in computer-mediated communication (CMC) brought not only new business opportunities for impendent fucking-machine inventors, they also proliferated new ways of having sex. It's easy to find open source sex sites everywhere on the Internet. *Smartstim* is a site for sharing electro-stimulation programming for sexual pleasure. *Cybermistress* is a site where Internet users can build and share their customized programming for a virtual mistress online. Among all of these sites, *Slashdong* and *Opendildonics* are two that advocate teledildonics productions based on community and public collaboration. Kyle Machulis, the owner of both sites created his first teledildonics 'Sex Box' by hooking up an Xbox controller with a dildo. He explained that his original inspiration for using video games was the fact that video games have replicated an imaginary world where people can do whatever they cannot do in reality. He said at Arse Elektronica 2007, 'there are communities of people with fetishes and they find them in video games… People play video games to follow up on what's in their imagination'. He posited a link between his creation and J.G. Ballard's car crash fetishism dealt with in his 1973 novel *Crash*. While people use his 'Sex Box' with the car racing game 'Crash', they can experience the 'reality' of car crash right on the screen accompanied by an orgasm. As a full-time robotic engineer, Machulis invented the 'Sex Box' out of his own curiosity. He then started to post his creations on his blog *Slashdong* and received 60,000 unit hits the next day. After three years of explorations on teledildonics, he became the expert in the open source teledildonics community. During my interview with Machulis, he said,

'I'm sharing all the information on the Internet, because I feel that people should build whatever they want and use it however they want. I'm not really worried about getting money out of it. I want the Internet users to take the instructions and apply them on their own fetish. Thanks to the internet now, I cannot keep up with those new fetishes anymore'
(Machulis, personal interview, 2008).

When Machulis was asked whether he had always been interested in sex before making the teledildonics, he said his focus was always on technology. He grew up in the mid-west in the United States and has been using the Internet since the mid-80s. He believes the Internet profoundly influenced his social growth and how he perceived things. In fact, he had not heard of the term 'teledildonics' and had no interest in sex machines before he became an expert in the field. What he had was an enormous passion for building things to satisfy his fantasies along with an interest in communicating via the Internet. He said during the interview, 'Since I spent most of my formative years talking to people on the Internet versus talking to the real people, it seems to make a lot of sense this way. Though, funny enough, I have a fiancé and I don't really use teledildonics in my personal life'. Instead of being preoccupied by sex, he explained that he had fetishes about engineering which he found very sexy. During the process of putting things together as an engineer, he always ignores the end goals while indulging himself in the experimentation. Like the 'Sex Box' project, he got pleasures by testing different video games and dildos to see what would happen. He never expected that eventually many online users would utilize his device. Undoubtedly, Machulis grew up as part of the Internet generation. Unlike many digital entrepreneurs, he rather keeps his sex and technological creations free, open and easily available. 'I just don't have any interests in monetizing the sex stuffs, because once you apply money on that, things become not fun', says Machulis. He explained his rationale for open sourcing teledildonics was that he wanted to do viral education on robotic engineering. He thought if sexual elements in advertising were aimed at making people buy things, his focus on sex was aimed at making people learn. He found it amazing that once people knew that the end goal of learning engineering was orgasm, they were automatically interested in broadening their knowledge. Instead of just providing programming for the Internet users to download, he explained every detail of how teledildonics worked on his websites and how he hoped to promote engineering technologies. The users of his websites are the tech-savvy generation; as the Internet became more and more ubiquitous, he expected that people from different generations and background would be interested in using his inventions. One of his recent experiments involved applying the biometric engineering technology to his sexual inventions. He wanted to map the organic idea of human reproduction into the computer space that 'doesn't need to reproduce'. He wanted to build sex machines that can provide synesthetic experience according to human biological data.

The technique that is being applied to open source teledildonics is computer engineering. Obviously, as an independent teledildonics producer, Machulis was fascinated by the process of designing software that was aimed at integrating sex with different hardware. The applications of computer engineering technology on sex were a vehicle for him to achieve his bigger visions. Just like Timothy Archibald and many other independent fucking-machines producers, the end result of producing projects on sex machines was not solely about exploring sexuality. Other than the functional nature of computer engineering, the 'technology' that was being 'fetishished' in the independent teledildonics production was the unpredictable 'high-techness' behind the source codes, i.e. the open source culture. Unlike the garage productions which rely on 'hands-on' tools and standard mechanics, the tools for independent teledildonics producers are the computer and the modem; the space for the teledildonics production is the 'blogopshere' instead of the automotive garage.

Blogopshere is a term that describes blogs, wikis and personal broadcasting that exist as a connected social network and community. The social and cultural space of the blogosphere is non-geographical and non-physical. Under this social network, like many other open source content providers of teledildonics, Machulis became the 'prosumer' by co-creating goods and services rather than only producing or consuming products. His creations were based on the existing hardware that was available in the market, such as dildos, Xbox and the video game 'Crash', though the products he created were uniquely different from the original products. The open source culture which motivates the appropriations, modifications and redistributions of products back to community or organizations was 'fetishized' for its inscrutable value and 'mysterious life'. The 'high-techness' of the open source culture of teledildonics allowed for a production and consumption that is not only based on how the body perceives physical pleasure, but how social networks and communities are magically involved with the idea and lifestyle of what sex and technology signify together. The do-it-yourself aesthetics, the process of discovery, the idea of achieving unpredictable results by altering existing products in the massive open social networks and the enjoyments of 'decentralizing' the conventional producer-generated (sex) products are what are being produced and consumed in the independent productions of teledildonics.

Re-articulation of gender dynamics in sex machines pornography

In *Sex Machines: photographs and interviews,* most of the fucking-machines were being produced by suburban men; women were mostly not interested in using the inventions. 'The recurrent thing in the book is the idea of men being really into these sex machines and women being puzzled by them. The women feel like they like hugging, cuddling and giggling, they don't want the sex machines. That is something more than a norm.' said by Timothy Archibald. For Kyle Machulis, the teledildonics he created were not being used personally with his fiancé. Many of the independent-made sex machines were not being designed for targeted users. The production of those machines was not aimed at achieving any practical end results. On the contrary, the industrial-made sex machines in the porn and sex toy industries were manufactured to make profits; companies made deliberate decisions on how to produce, market and distribute their sex machine products.

Repositioning the cultural phallus

Fucking Machine is one of the most successful porn sites which is wholly dedicated to human-machine interactions in pornography. During my interview with Thomas Roche, he explained why fucking-machine was a popular genre. He thought one of the reasons was that the effects of sex machines on people were much more direct and obvious for physical pleasure. The design of those fucking machines were about women in control of their pleasure from multiple modes of thrusting stimulations that a human being and a vibrator cannot provide, no matter if this interaction was on or off-camera. At the opening of *Arse Elektronika 07*, curator Johannes Grenzfurthner asked if any one in the audience would be willing to do a live demonstration with the fucking-machines. A university student called Binx, who didn't know anyone from the conference, volunteered to be penetrated on stage by 'Fuckzilla', a fucking-machine that was produced by *Kink.com*. She wanted to experience the feeling of being penetrated by a high-powered fucking-machine in front of total strangers and to find out her subjectivity as a woman while being on stage.

> 'I volunteered to get on stage and fuck a $10,000 machine, undoubtedly a once-in-a-life-time opportunity. Thrill-seeker? Exhibitionist? Robot-fetishist? Yes, yes and yes…Perched naked on a table behind a sheet, I admit that I started off a little nervous. All of my trepidation fell aside, however, once the actual show started. The Fuckzilla made it hard for me feel anything but Intense Pleasure, oh yes, with capital letters. I had the hands-down best orgasm of my life, both subjectively and objectively… But, do you want to know the real reason I agreed to get on stage with Fuckzilla? Feminism, baby', said by Binx (Grenzfurthner, pp,83, 2008).

Reports said that Binx had 'squirted about five feet into the air', and Binx described that the penetration was a force she had 'never came close to achieving'. Obviously, the intensity of physical pleasure and sexual climax from the penetration by 'Fuckzilla' was something that Binx had never encountered before. The fact that Binx's actions were being displayed in public and broadcast on the internet like a pornographic actress, made her think about what she had learned about feminism theory and gender study in the university. She thought fucking-machines were 'the pornographic equivalent of third-wave feminism', that she was able to proclaim her own sexuality and self-conscious empowerment by not faking her orgasm for the male gaze. 'Fuckzilla' was a remodeled 'Johnny 5'; its arms were mounted with a variety of sex toys including synthetic silicon tongues that were able to move up and down. Binx described that the machines are designed to 'get women off, nothing more and nothing less', and the fucking machines porn was a 'fundamental shift towards the woman's enjoyment in the total absence of men'. She claimed that being independent of men during sexual intercourse and the willingness to try new things went in line with her sex-positive feminist principles.

To examine Binx's reflections in reaction to 'Fuckzilla', one needs to understand the dynamics between Binx and the giant high-powered 'phallic' object. In 'The Lesbian Phallus', Judith Butler described 'phallus' as 'transferable, substitutable, plastic, and the eroticism produced within such an exchange depends on the displacement from traditional Musculinist contexts as well as the critical redeployment of its central figures of power'. Based on Lacan's 'phallus' that displaces the male genital organ from its ontological reality, Butler argues that both man and woman can 'have' and 'be' the 'phallus' symbolically (Butler, pp 85, 1993). She considers that 'having' the phallus can be symbolized by any body part or 'purposefully instrumentalized body-like things', and that the 'signifying chain' of 'having' and 'being' of the 'phallus' can be recirculated and reprivileged from the 'logic of non-contradiction that serves the either-or of normative heterosexual exchange' (Butler, pp88, 1993). Butler displaces the Lacanian formulation of 'phallus' which implies that the signifier of phallus is being performed by women in which the process of 'self-definition and 'potential autonomy' are being excluded (Grosz, pp116, 1990). The 'signifying chain' of 'Phallus' in this sense doesn't only belong to men, and thus when women are performing penetrative sex with phallic objects they are not necessarily representing what men's biological penis or symbolic phallus can perform during sex act. On one level, the machine that Binx used was biologically incapable of ejaculation. They were the pleasure-giver designed solely to assist women's autonomous desires. On the other level, in Butler's conceptualization of phallus, Binx (biological woman without a penis) was 'having' the symbolized phallus ('purposefully instrumentalized body-like things') that was symbolically 'reterritorialized' and 'subverted' from the biological penis (men). The idea of 'phallus', both the physical functions and symbolic significations were displaced from the 'the either-or of normative heterosexual exchange' that reinforces the male to female 'orgasm for orgasm' kind of exchange. The high-powered fucking-machine symbolizes a 'force (that) she (Binx) had never come close to achieving before. The force that Binx 'purposefully' encountered reinforced the autonomous ideal of the third-wave (post-) feminism. The symbolic phallus in 'Fuckzilla' did not belong to anyone and everyone; it was being internalized by Binx as a symbol that is 'displaceable', 'performative', and even 'phantasmatic' through chains of imagination along with the absence of a biological phallus and/or a symbolic male phallus during penetrative sex.

Correspondingly, as noted by Thomas Roche, most of the new talents in the porn industry who applied to *Kink.com* found fucking-machines appealing because interacting with a machine was perceived as less of a social stigma than performing with a man or even with another woman. While Binx thought that interacting with the fucking-machines was empowering for her, Roche thought his productions were liberating for male consumers as well. The majority of customers of *Fucking Machine* are men, men who enjoy watching woman who really want to get penetrated and truly enjoy themselves for the autonomous straight physical pleasure. Just as Timothy Archibald stated, Roche believed that there was a cultural message and stereotype that in many societies women were preoccupied by the idea of intimacy such as cuddling and pillow talk before and after sex. By seeing some women who openly wanted hard thrusting and demanded a machine to give it to them was also liberating and stereotype breaker for men. He added, 'The idea of sexually is an option, it's not something that is dictated by the biological or social positions you hold. It's a choice that you make in order to enjoy yourself.'

Undoubtedly, the design of fucking-machines in pornography was aimed at creating visual pleasure (mostly for men) (Mulvey, pp.19, 1989). The female gender being portrayed in *Fucking Machine* was not politically and culturally neutral; they were being mediated within the patriarchal system, deep-seated in society and the pornography industry. This paper is not aimed at proving that the productions of fucking-machines in pornography are completely detached from the socially and culturally constructed gender norms, nor to prove that pornography is only and essentially reinforcing women's oppression, though during the process of the productions of fucking-machines in pornography, the socially constructed female has been re-articulated symbolically. The sexual acts of the pornographic actress were being displayed for pleasure and was autonomously and deliberately executed and controlled without the mediation by any director. The fucking-machines were being controlled and adjusted simultaneously in response to the actress's physical and mental self-reactions. The emphasis on real orgasm, female ejaculations and actress's desire in *Fucking Machine* was a contrast to the mainstream pornography that emphasized on male's desire and ejaculations (on female bodies). Fucking-machines were being articulated as a mediator for sexual pleasure that doesn't include a biological male organ and symbolic male desire. Fucking-machines repositioned the female from the passive to active role, and the 'male/active' social construction was being deemphasized due to the lack of a biological and symbolic male. The fantasy being projected in the fucking-machines porn was shifted from a totally male-oriented pleasure-seeking body to a pleasure-giving body (machine).

Mediated-voyeurism of the teledildonics porn

While the majority of consumers of *Fucking Machine* are male for one of the biggest custom-made sex machine companies, Thrill Hammer, the larger demographic is women. Ranging from the gynecological chair with a high-powered dildo that is networked with the Internet, to an aromatherapeutic fucking machines that is equipped with vaporizer, Allen Stein's products are primarily targeted to women clients. During my interview, Stein said that the key market for his custom-made fucking-machines was usually wealthy individual women who were making six digit annual incomes. He said,

'They are usually the professionals, such as doctors and lawyers, people who have the busy schedules and don't have time to date. On Thrill Hammer, you can't fake the orgasms; it was so effective that it pops a lot of orgasms, a lot of girls who haven't had the first orgasm with it before' (Stein, personal interview, 2008).

Stein believed that many of his clients were not satisfied with the social norms of dating. Instead of having sex machines as a total replacement for men, they used sex machines as an option to fulfill the momentary desires for penetrative sex. Apart from fucking-machines, Stein is the first producer in the industry of custom made teledildonics which applies the teledildonics technology to pornography. After being in the pornography industry for five years, Stein noted the excess of pornography in the market that was faking women's orgasm and sexuality. He claimed that his teledildonics, which were being documented with *Sex Machine Cams*, was a breakthrough in technology and women's sexuality. *Sex Machine Cams* is the first commercial porn site that broadcasts pornographic actresses and actors interacting with sex machines that are controlled by Internet users in real time. The virtual set allowed real time video and audio streaming for the Internet users to realistically communicate with the performers. Stein thinks that the virtual sex machines provide another level of intimacy for his customers *and* the performers, he described the users of the site as sensual; they were there to pleasure women instead of just sexually dominating them. At the website, users can ask the porn stars if they want the machine to be faster or slower and if they want more vibrations or rotations. Users can understand how performers feel through their actual verbal response. Stein believed that those feedbacks provide a location for better sexual experience, as this kind of communication rarely happens during bodily sex and in conventional pornography. Stein said, 'If guys learn how to slow down and communicate, there wouldn't be pre-mature ejaculation, there wouldn't be all these gender imbalances in sex'.

If women in traditional pornography are the objects of men's voyeuristic gaze, pornography that is transmitted via webcams is a 'mediated voyeurism' (Senft). Among all the current theorizations of gazing via webcams, Teresa Senft analyzed that webcams sex operated through 'an aesthetic of the 'grab' ', it provided capabilities for viewers to 'take what they see in bits and pieces, out of sequence, and to re-make it according to their own desires'. She argued this phenomenon from a feminist perspective that performers on webcams can be 'grabbed by Web consumers as a self-branded 'super cyborg'' and they were 'capable of resolving all contradictions about women in network society'. On *Sex Machine Cams*, both the 'voyeur' and the 'exhibitor' are mediating the 'grab' of imagery. On one hand, spectators pay to control the teledildonics by the knobs and verbally instruct the performer by using a microphone; the performers can easily be objectified and made visible on the computer screen in order to entertain the user's gaze. On the other hand, the performers own the TriCaster with the ability to 'create, broadcast, web stream and project' (newtek.com) what and how to be visible on the spectator's screen which can creates new dynamics of representations of female bodies in pornography. During the 45 minutes of live demonstrations of the performer Summer with her client on Oct 3rd 2008 at the studio of Sex Machine Cams, it was observed that the 'spectatorship' initiated by the user was being completed and resisted at the same time, no matter if it was deliberately informed by Summer or by the ontological nature of the system. Summer accepted and rejected the user's request through flirtatious negotiations; the tactics became convincingly erotic by the use of mis-en-scene or even the pixels, flickers and disconnections that were being rendered by the machine. The fractual 'positioned view' rather than a 'unified perspective' being mediated in webcam pornography (Grenzfurthner, pp.166, 2008) made the 'self-branded 'super cyborg'' able to re-articulate her objectified female identity in the networked society. The position of the performer was symbolically higher than the user, since she was mastering the illusions and fantasies with the broadcasting system, deliberately and impulsively working it at the same time. The 'grabbing' that was being mediated by the 'voyeur' and 'exhibitor' via the webcam technology destabilized and intervened with the 'visual pleasure' that is usually articulated on conventional pornographic screens.

From biological body to the transformation of new bodies

The body of fucking-machines and teledildonics represent the 'modernist machine aesthetics' of functional and instrumental ideals (Rutsky, pp12, 1999), while humanoid sex machines were designed to resemble the biological human structure. In *Turing's Man*, J. David Bolter states, 'there was perhaps never a moment in the ancient or modern history of Europe when someone was not pursuing the idea of making a human being by other than the ordinary reproductive means'. The desire to recreate a human body through artificial means can be noted throughout history.

First Androids, located at Nuremberg, Germany, is a company that produces custom-made humanoid sex dolls that can be hooked up with machines and enact different movements for sexual stimulation. Michael Harriman, the owner of First Androids identified himself as an artist rather than a businessman. He told me that the sex dolls were being made in various forms and with various functions upon the specific requests of his clients. According to Harriman, his clients ranged from teachers to police to priests and came from all over the world, from Germany to China. After filling out the First Androids online forms and sending pictures of the desired type, clients can explain every detail to Harriman concerning their particular desires. The products are then mailed to the clients from 2 weeks to a year's time, depending on the complexity of the doll requested. Along with the conventional male and female dolls, some of the First Android's dolls are called 'fantasy sculptures' which come in irregular or mutated body forms. He said during the interview,

> 'I made one look like a bed with a lot of breasts, holes and mouths. I also have the other doll with 6 breasts and a huge vagina-like hole. Customers have their own fantasies, and I try to fulfill them. I also made a breast wall for one of my clients. And of course I made many transgender dolls, these dolls are better than human because they have many different sexual features' (Harriman, personal interview, 2008).

Underneath the silicon skin, some of the dolls are equipped with internal heaters and electronic hearts that are able to beat faster during the 'sex' act. The dolls are capable of mimicking suggestive body moments while being remotely controlled by the user. Harriman's ultimate fantasy is to create a 'Cherry', an artificial woman that is programmed to be the ultimate wife and erotic companion in *Cherry 2000*, a science fiction cult film that portrays a perfect humanoid sex machine. He thought the perfect sex doll should have the capability to do house work along with sex service. In fact, many users were not only consuming First Androids' products solely as sexual devices, but as Harriman said,

> 'A lot of customers have their dolls sitting on their chairs or lying in their bed all the time. Some of my clients even have their dolls sit with them in their car for companionship. They live with the dolls, and treat them as humans. They put them in a garden and take a sunbath together. People have been interested in synthetic human for thousands of years. It's the dream of the mankind to make a copy of himself' (Harriman, personal interview, 2008).

To understand what it means to fantasize and design the humanoid sex machines and the dynamic relationships between the producer, consumer and the sexualized humanoid body, one needs to understand how our body is being culturally interpreted. In Baudrillard's view, our body is a 'marked' body that is organized in the 'system of signification' of cultural codes. Our body is being articulated by the mediation of 'directive models and thus under the control of meaning', the process of such is the 'transference of the wish-fulfillment of desire upon the code.' Throughout history, humans have sought ways to re-articulate our bodies, from cosmetic aids to plastic surgery, from medicine to surgically implanted devices. Our body has been intervened with in many different ways in order to go beyond its ontological limits. The imperfect, fragile, and mortal human body is thus being manipulated to fulfill our desire and to signify the socially and culturally constructed codes of the ideal body type for ourselves or for the physical partner. These manipulations are not only being applied to our biological self; our body has additionally been dreamed about or projected onto the 'prosthesized' form. In *Simulacra and Simulations*, Baudriliard writes,

> 'Everyone can dream, and must have dreamed his whole life, of a perfect duplication or multiplication of his being, but such copies only have the power of dreams, and are destroyed when one attempts to force the dream into the real… The same is true of the (primal) scene of seduction: it only functions when it is phantasmed, reremembered, never real. It belonged to our era to wish to exorcise this phantasm like the others' (Baudrillard, pp.95, 1994).

Sex dolls are being created to resemble our human bodies, complete with texture, color, shape, temperature, seductive gestures, flexible genitals and even delicate respiratory prosthetic lung designed and crafted to satisfy the users' 'wish-fulfillment of desire' that is coded by the social and cultural ideals of human bodies. Unlike our biological body, which is guided by material limitations, the 'prosthesized' bodies can be made in any fantasized forms that multiply different codes of signs (simulacra) through the 'power of dreams'. In reality, our natural bodies are being 'marked' with the fact that it is constantly being bought, sold, exchanged, replaced, trained, treated, examined and designed. Our signified body is also constantly being integrated and disintegrated. The fantasy dolls with multiple genitals and irregular body shapes create a mirror image of the fragmented and multiplied signs of our real body through the 'power of dreams'. The potentially unlimited possibilities of body manipulation with the assistance of technology allow us to dream limitedly about the 'perfect' self and the other (self). During my interview with Michael Harriman, he said that his ultimate dream was to create a Cherry; in fact, many of the consumers treated their dolls as their wife, lovers, and domestic companions. The intimate relationships between the consumers and the sex dolls well represent the state of 'phantasm' that is being processed between the self and the other (self). According to Baudrillard, 'phantasm' is capable of creating 'strangeness, and at the same time the intimacy of the subject to itself are played out' (Baudrillard, pp95, 1994). The 'phantasization' of the sex dolls heightens the pleasure and intimacy that is mediated by the 'power of dream' of the 'prosthesized' body. The lack of realness of the non-biological sex dolls provides room for the users to dream or even transgress what is considered socially and culturally acceptable with a real lover since the 'prosthesized' bodies will never reject their owner's wishes. The 'prosthesized' bodies are never going to be real biologically, and paradoxically, they are forever going to be real in the realm of 'phantasization'. The signified dream wife, lovers, or domestic companions are being 'phantasized', produced and consumed by the producers and consumers of the sex dolls. The human body is being re-articulated symbolically through the imagination of the other body; pleasure and intimacy are being processed and transformed during the production and consumptions of the humanoid-like sex machines.

These kinds of humanoid sex machines are not only industrially produced as a sex doll, the 'phantasm' of sex dolls and robots are concomitantly being culturally produced as a discourse on human fidelity. The idea of 'phantasm' between the human owner and the robots has been much discussed in recent years among academics. Peter Asaro, a scholar and director of the documentary *Love Machine*, said at *Arse Elektronika 2007* that many discussions on human and robotic relationships were focusing on the idea of fidelity. In his view, people who are in love with sex dolls are formulating their very own definition of fidelity, while technologically the dolls have become the other, a foreign being that is not an extension of the self, but actually another being. Similar discussions are being raised by David Levy, the author of *Love + Sex with Robots* who predicted robotic marriage will be legalized by 2050 in the United States. Levy says 'Keeping a robot for sex could reduce human prostitution and the problems that come with it. However, in a marriage or other relationship, one partner could become jealous or consider it infidelity if the other uses a robot' (Choi, 2007). The imagination of the other body in reactions to sex robots has raised questions about the socially and culturally constructed borders of not only sex, but love and intimacy. The 'phantasm' between human and 'prophesized bodies' is being translated as a cultural product, the discourse of the possibility of a new kind of love and sexuality.

The end of body - the bioengineered humanoid sex machines

When does 'phantasm' become real? Baudrillard says 'the power of dream' would be destroyed when the dream is being forced to be real; when 'phantasm' is being realized and materialized into 'flesh and bone' ', one is changing 'the game of the double from a subtle exchange of death with the Other into the eternity of the Same'.

In recent years, the advancements in biotechnology, a technological application that modifies biological organisms, has made genome alterations and reproductive cloning possible. The day when a replica of the human can be artificially reproduced is no longer completely a science fictional dream. Before our era, when an actualized cloned sex humanoid body may be possible, the dream of having a sex replica was portrayed in such films as *I.K.U.* Directed by Shu Lea Cheang, *I.K.U.* is an experimental pornographic film that portrays a new form of sex machines that have been invented by a futuristic corporation. The 'prosthesized' in *I.K.U.* is a new bioengineered organism that is made with the advanced cloning technologies developed by GENOM Inc., a Japanese multinational corporation that handles information technology and genomes research. In this imaginary world, consumers can download and experience orgasms without any physical contact solely through the power of their brain by using an orgasm decoder chip called the 'I.K.U. Chip' which is sold in vending machines. During my interview with Cheang, she said that she has been always interested in exploring the philosophical questions about the bioengineering and cloning of the human body in her artwork. She sees her work as addressing the 'fear of cloning and all the big concerns about creation'. The new sexual organism in I.K.U. was created to ask the question, 'what will the human race create from the new pleasures gained after being free from the ecstasy gained by physical friction?' (i-k-u.com).

In the case of First Androids productions, sex dolls are being produced and projected as 'the other', to be eroticized. The material body of the sex dolls is the mirror of the symbolic simulated human body. In *I.K.U.*, the bioengineered organism is not the reflection of human body, in fact, the body of the sex robots is biologically and symbolically identical to a human body; the 'immutable repetition of the prosthesis', and the 'power of dream' is superseded. In Baudrillard's sense, the cloned body 'negates the subject and the object', it is only the 'micromolecular genetics (that) is nothing but the logical consequence'. This 'prosthesis par excellence' also symbolizes the end of body, Baudrillard writes,

> 'But when one reaches a point of no return (deadend) in simulation, that is to say when the prosthesis goes deeper, is interiorized in, infiltrates the anonymous and micro-molecular heart of the body, as soon as it is imposed on the body itself as the 'original' model, burning all the previous symbolic circuits, the only possible body being the immutable repetition of the prosthesis, then it is the end of the body, of its history, and of its vicissitudes… Cloning is thus the last stage of the history and modeling of the body, the one at which, reduced to its abstract and genetic formula, the individual is destined to serial propagation' (Baudrillard, pp. 100, 1994).

The 'end of body' in Baudrillard's sense can be interpreted as the end of the 'other' body. Without the difference between the subject and the object, the 'other' is forever identical to the self, the 'system of signification' of the 'marked' body is forever to be repeated and thus effaced. Without the 'power of dream', there will be no prosthetic pleasure, and therefore no intimacy and strangeness between the 'prosthesized' bodies and the self (if the 'self' can still be identified). The mortal and undifferentiated body is a pure material body that is filled with genetic data but nothing else. The end of intimacy and interaction between the 'prosthesized' body and the self is well represented by Cheang's idea of sex free from 'physical fictions'. In *I.K.U.*, physical sex is no longer necessary; sex is reduced to biotechnological data that can be bought from a vending machine. Instead of having the prosthetic bodily pleasure, the pleasure of sex in *I.K.U.* is a pure consciousness kind of pleasure. Baudrillard identifies the first trends of sexual liberation in the 60s as dissociation from procreative sexual activity. He thinks that we're now in the second phase, which is the dissociation of reproduction from sex. He says, 'Among the clones, sex, as a result of this automatic means of reproduction, becomes extraneous, a useless function' (Baudrillard, pp.10, 2000). In the future imagination of humanoid sex machines, the body is being rearticulated back to a pure material substance, and pleasure is redefined as pure consciousness. The immortal body of the other and the self are bound up in a state of obsolescence where the material body (functions) become meaningless and yet curable, replicable and replaceable.

A brave new world - new configuration of sex, technology, gender and body

In the field of independent, industrial, artistic and institutional productions of sex machines, each agent has his or her own formulations of machines, materiality and human sexuality. Values, beliefs, ideologies and practices upheld by different agents are being processed and circulated within the sexual field and this has rendered the sex machine a cultural artefact.

The techniques being applied to the independent sex machine productions have inspired the porn industry to utilize these products as another way to package the orgasm. The success of industrial sexual products has, in return, motivated independent inventors. The science fictional imaginations of future sex machines have become the signified ideals for sex dolls makers and consumers. The new pleasures and sensations that are being produced within the human/dolls relationships have brought up new discourses of sex and fidelity in popular films and academic conferences. Different agents within the field of sex machine production have created a semiotic network that is constantly defining and re-defining how we make sense of and utilize sex machines materially and culturally. The three types of sex machines being technically produced not only engender different kinds of pleasure and sensation, but the process and result of production also encode how sex machines are made culturally meaningful. The cultural meaning of sex machines are being articulated by different contingencies of circumstances (Hall, pp.3, 1997) and as a consequence, alter the role of sexuality, technology, body and gender into different forms.

Sex-positive culture has detached sex from procreative norms. The insertion of sexuality into technology in the field of sex machines has further posited new sensations and new meanings of sex. Sex machines have not only brought new kinds of physiological pleasures to many users, the productions of sex machines have also created new forms of sexual strangeness, otherness and identity. The 'other' body of sex machines can range from the networked vibrator such as Xbox, 'Johnny 5' to the undefined humanoids that are being used as sexual and/or love partners. Unlike the previously made sex toys, many users and producers are treating sex machines not only as a material object or device that is used in facilitating human sexual pleasure, they also consider and project sex machines as the 'other', sexually and spiritually. These machines are being interpreted by different producers and consumers as something strange, something in between a human and sex toys, something that doesn't have a concrete identity or even a name. Sexuality has thus been displaced even further from our fragile understanding of sexual orientations and practices, since the sexual 'other' in the sex machine engenders complex and fragmented significations of identity and practice. The previously defined terms for sexual identity such as 'heterosexuality' 'homosexuality', 'bisexually', 'asexuality', 'polysexuality', or the terms for non-normative practices such as sexual fetishism are thus being further reinterpreted and rearticulated. Is the sexual 'other' a lover? Are they gendered? Are they only a toy? Can they substitute for human bodies? Can they substitute spiritually for the human? What if they become bio-engineered and what if they have intelligence?

The strangeness and 'otherness' of sex machines destabilize how we understand our sexual identities. The meaning of sex machines is constantly being constructed and deconstructed by different agents within the sexual field. Until the day we have the cloned body for sex, this sexualized technology in the realm of sex machines will always be considered as a foreign object for producers and users to share, to live and to be with physically, sexually and (or) spiritually. This kind of strangeness and 'otherness' that is generated from the intersection of sex and technology has also reconfigured body and gender roles. While still having intense and unique bodily sensations, the human body has been further disintegrated and displaced by sex machines. The biological human body is being replaced by a foreign other, the pornographic body is being negotiated and rendered via teledildonics, the signified human body can be further and unlimitedly 'prophesized' and 'phantasized' all the way to the end. The non-essential genders are further problematized by the 'other's' gender that is transferable, substitutable and undefinable. Since the biological gender of the other is absent, the signified gender of such is being reshuffled and reprocessed by the producers' and users' imaginations.

What exactly is this something-in-between 'other' being created during the production of sex machines? What does this strangeness signify? Before the day of doubling 'the Other into the eternity of the Same', sex, technology, body and gender will always be configured as new possibilities within the realm of sex machines innovations.

References

Journals

Green, Adam Isaiah. The Social Organization of Desire: The Sexual Fields Approach. Sociological Theory 26 (2008): 25-50.

Miller, Laura J.. Family Togetherness and the Suburban Ideal. Sociological Forum, Vol. I0, No. 3 (1995) 393-418.

Senft. Teresa. Camgirls: Webcams, LiveJournals and the Personal as Political in the age of the Global Brand. Ph. D. Thesis, New York University, New York, NY.

Books

Archibald, Timothy. Sex Machines: Photographs and Interviews. Los Angeles: Daniel 13/Processm, 2005.

Levy, David. The Evolution of Human-Robot Relationships: Love + Sex with Robots. New York: HarperCollins Publishers, 2007.

Mauss, Marcel and Schlanger, Nathan. Techniques, Technology and Civilisation. New York: Berghahn Books, 2006.

Kellner, Douglas. Baudrillard: A Critical Reader. Blackwell Publishing, 1994.

20 Rutsky, R. L. High Techne: Art and Technology from the Machine Aestetic to the Posthuman. Minneapolis: University of Minnesota Press, 1999.

Berger, Peter and Luckmann, Thomas. The Social Construction of Reality: A Treatise in the Sociology of Knowledge. Michigan: Penguin, 1967.

Grenzfurthner Johannes, Günther Friesinger, Daniel Fabry (eds.). Pronnovation? Pornography and Technological Innovation. San Francisco: Re/search Publications, 2008.

Butler, Judith. Bodies that Matter. New York: Routledge, 1993.

Grosz, Elizabeth A. Jacques Lacan: A Feminist Introduction. Routledge, 1990.

Mulvey, Laura. Visual and Other Pleasures. Indiana Univ Pr, 1989.

Baudrillard, Jean. Symbolic Exchange and Death. SAGE, 1993.

Baudrillard, Jean. Simulacra and Simulation. University of Michigan Press, 1994.

Baudrillard, Jean. The vital illusion. Columbia University Press, 2000.

Hall, Stuart. Doing Cultural Studies: The Story of the Sony Walkman. SAGE, 1997

Interviews

Archibald, Timothy. Personal interview. 27 Sept 2008.

Machulis, Kyle. Personal interview. 29 Sept 2008.

Levy, David. Personal interview. 17 Aug 2008.

Roche, Thomas. Personal interview. 30 Sept 2008.

Stein, Alan. Personal interview. 3 Oct 2008.

Harriman, Michael. Personal interview, 7 Aug 2008.

Cheang, Shulea. Personal interview. 20 July. 2008.

Websites

'Slashdong'. Machulis, Kyle. 2005-2007. NP Network. 3 Jan, 2009. < http://slashdong.org>

'Opendildonics'. Machulis, Kyle. 24 November 2007. Wikipedia. 3 Jan 2009. < http://wiki.opendildonics.org>

'MMOrgy '.Machulis, Kyle. 28 May 2007. MMOrgy. 3 Jan 2009. < http://www.mmorgy.com>

'Fucking Machines', Peter Acworth. 1997-2009. Kink.com. 3 Jan 2009. < http://fuckingmachines.com>

'Sex Machine Cams'Alan Stein. 2008. Butter Butter Productions Inc. 3 Jan 2009. < http://sexmachinecams.com>

'The Thrill Hammer', Alan Stein. 2006. Thethrillhammer.com. 3 Jan 2009. < http://dnn.thethrillhammer.com>

'First Androids'. Michael Harriman. 2007. First Androids. 3 Jan 2009. < http://www.first-androids.org>

'Kink.com' Peter Acworth. 1997-2009. Kink.com. 3 Jan 2009. < http://kink.com>

'TriCaster Portable Live Production' 2009. NewTek. 3 Jan 2009. <http://www.newtek.com/tricaster>

'Sex and marriage with robots? It could happen' Choi, Charles Q. 12 Oct 2007. Microsoft. 3 Jan 2009. < http://www.msnbc.msn.com/id/21271545>

'Introduction' Cheang, Shulea. 3 Jan 2009. < http://www.i-k-u.com/eng_h/iku/int_j.html>

Films

I.K.U. Dir. Shu Lea Cheang. Uplink Co., 2001.

Cherry 2000. Dir. Steve De Jarnatt. Orion Pictures, 1987.

Love Machine. Dir. Peter Asaro. Kaiczech & Savario Productions. 2001.

Jason Brown

MIND DIDDLERS

Happy Trails

In 1929 William S Burroughs was sent to Los Alamos Ranch School in New Mexico. Upon arrival, each boy was assigned a horse, and they were divided into patrols (Piñon, Juniper, Fir and Spruce) by size and ability rather than age. While at Los Alamost, Burroughs had his first sexual experience with another boy at the school.

From the Los Alamos Ranch School anthem:
> *Hio, we sing of the happy days*
> *Hio, we sing of the days of joy*
> *Winter days as we skim o'er the ice and snow*
> *Summer days when the balsam breezes blow*
> *Los Alamos!*

Burroughs left Los Alamos in 1930 (the year J G Ballard was born). Perhaps he was expelled after taking chloral hydrate with another student. Or perhaps he simply persuaded his family to let him return to St. Louis.

In November 1942, the Los Alamos Ranch School was purchased by the United States Army's Manhattan Engineering District, who took control of the property in February 1943. The facility was first known as 'Site Y' and eventually as Los Alamos National Laboratory. The buildings where Burroughs first explored his sexuality were used by the scientists and engineers to design the world's first atomic weapon.

Agape

In 1939, Jack Parsons and his wife Helen joined the Agape Lodge of Crowley's O.T.O.
Agape refers to spiritual or 'higher' love as opposed to Eros, the 'lower' or sexual love.

Rim Job

[A] distinction has to be introduced here between two types of lack, the lack proper and hole: lack is spatial, designating a void *within* a space, while hole is more radical, it designates the point at which this spatial order itself breaks down (as in the 'black hole' in physics). Therein resides the difference between desire and drive: desire is grounded in its constitutive lack, while drive circulates around a hole, a gap in the order of being. In other words, the circular movement of drive obeys the weird logic of the curved space in which the shortest distance between the two points is not a straight line, but a curve: drive 'knows' that the shortest way to attain its aim is to circulate around its goal-object. (One should bear in mind here Lacan's well-known distinction between the aim and the goal of drive: while the goal is the object around which drive circulates, its (true) aim is the endless continuation of this circulation as such.)
- Slavoj Žižek

Desiring Machines

'It might be a Scottish name, taken from a story about two men in a train. One man says, 'What's that package up there in the baggage rack?' And the other answers, 'Oh that's a MacGuffin.' The first one asks, 'What's a MacGuffin?' 'Well,' the other man says, 'It's an apparatus for trapping lions in the Scottish Highlands.' The first man says, 'But there are no lions in the Scottish Highlands,' and the other one answers 'Well, then that's no MacGuffin!' So you see, a MacGuffin is nothing at all.'
- Alfred Hitchcock

What we have here is a failure to communicate

From the magickal diary of Jane Wolfe, December 1940:
'Unknown to me, John Whiteside Parsons, a newcomer, began astral travels. This knowledge decided Regina to undertake similar work. All of which I learned after making my own decision. So the time must be propitious.
Incidentally, I take Jack Parsons to be the child who 'shall behold them all' (the mysteries hidden therein. ALI, 54-5).
26 years of age, 6'2', vital, potentially bisexual at the very least, University of the State of California and Cal Tech., now engaged in Cal. Tech. chemical labratories developing 'bigger and better' explosives for Uncle Sam. Travels under sealed orders from the government. Writes poetry - 'sensuous only', he says. Lover of music, which he seems to know throughly. I see him as the real successor of Therion. Passionate; and has made the vilest analyses result in a species of exaltation after the event. Has had mystical experiences which gave him a sense of equality all round, although he is hierarchical in feeling and in the established order.'

From a letter by Crowley to Wilfred Talbot Smith, undated, postmarked November 3 1943 (from The Unknown God, Martin P. Starr) :
'Apart from all else, your sexual acrobatics tended to give the Order the reputation of being that slimy abomination, a 'love cult.' Already in 1915 in Vancouver, all I knew of you was that you were running a mother and her daughter in a double harness. Since then, one scandal followed another.
'Your attempts to seduce newly-initiated women by telling them that you were now in a position to order them to sleep with you were acts of despicable blackguardism. What grosser violation of the Law of Thelema can one imagine? Not to mention that by English law, you might if successful have been found guilty of rape, and I should have heartily approved a sentence of penal servitude.'

From a letter by Jack Parsons to Alesiter Crowley regarding his first meeting with Marjorie Cameron:
'The feeling of tension and unease continued for four days. Then on January 18 [1946] at sunset, whilst the Scribe and I were on the Mojave Desert, the feeling of tension suddenly stopped. I turned to him and said 'it is done', in absolute certainty that the Operation was accomplished. I returned home, and found a young woman [Marjorie Cameron] answering the requirements waiting for me. She is describable as an air of fire type with bronze red hair, fiery and subtle, determined and obstinate, sincere and perverse, with extraordinary personality, talent and intelligence. During the period of January 19 to February 27 I invoked the Goddess Babalon with the aid of magical partner [L Ron Hubbard], as was proper to one of my grade.'

Crossing the Abyss

On 10 August 1946, L Ron Hubbard and Sara Northrup were married. Hubbard was still married to his first wife, Margaret Grubb.

On 1 December 1947, Aleister Crowley died of a respiratory infection in a boarding house in Hastings. His doctor died within the day. Some claimed that the doctor had refused to continue Crowley's opiate prescription so Crowley cursed him.

On 23 April 1951, Sara Northrup Hubbard sued L Ron for divorce on the basis of bigamy, torture, kidnapping, and Ron's demands that she commit suicide.

On 17 June 1952, Jack Parsons was killed in an explosion of fulminate of mercury at his home laboratory in Pasadena. His last words were reportedly: 'I'm not finished.'

Fallen

(Roswell...)

> *Hio, we sing of the youth time*
> *Hio, we sing of the open sky*
> *Ever true to Los Alamos we shall be*
> *Pledge anew forever our loyalty*
> *Los Alamos!*

Orthon

In 1949, George Adamski composed a science-fiction book - ghost written by Lucy McGinnis - entitled Pioneers of Space. At one point he submitted the manuscript to editor Ray Palmer.

On 20 November 1952, Adamski and several of his friends were in the Colorado Desert near the town of Desert Center, California...

In 1953, Adamski published a book he coauthored with Desmond Leslie, Flying Saucers Have Landed. This book marked the beginning of the modern alien contactee movement.

In 1954, Marjorie Cameron starred in Kenneth Anger's film Inauguration of the Pleasure Dome, which also featured Anais Nin and director Curtis Harrington.

In 1955, Adamski published Inside the Space Ships, ghost written by Charlotte Blodget. Though it was presented as a true story, Inside The Space Ships is a close re-write of the sci-fi novel Pioneers of Space.

Thanatos

Men have gained control over the forces of nature to such an extent that with their help they would have no difficulty in exterminating one another to the last man. They know this, and hence comes a large part of their current unrest, their unhappiness and their mood of anxiety. And now it is to be expected that the other of the two 'Heavenly Powers'... eternal Eros, will make an effort to reassert himself in the struggle with his equally immortal adversary. But who can foresee with what success and with what result?
- Sigmund Freud, Civilization and its Discontents (p.145)

Colonel Tom Edwards: Why, a particle of sunlight can't even be seen or measured.

Eros: Can you see or measure an atom? Yet you can *explode* one. A ray of sunlight is made up of *many* atoms!

Jeff Trent: So what if we *do* develop this Solar-thing-a-ma-call-it? We'd be even a stronger world power than now.

Eros: A stronger world power. A *stronger* world power! You see, you see! You're stupid! Stupid!

Jeff Trent: That's all I'm taking from you! [punches Eros in the face]

- Plan 9 from Outer Space

Bela Lugosi's Dead

In 1959, Naked Lunch and Plan 9 from Outer Space were first released. Both of these literary experiments are non-linear works on the fringe of science fiction dealing with horror, sex, death, and the alteration of consciousness, with extensive use of cut-up technique. But of the two, Plan 9 makes the most sincere gestures towards a plot: Aliens are trying to communicate with humans who studiously avoid paying attention to them, no matter what the aliens do. In a desperate bid for attention, the aliens initiate 'Plan 9' - the re-animation of dead bodies as a means of communication.

This question - how would we know non-human life forms were communicating to us - was not new.

In 1950, the physicist Enrico Fermi had a conversation with his colleagues at the Los Alamos National Laboratory while walking to lunch. Among other things, they discussed recent UFO reports and the possibility of alien life. Suddenly Fermi asked: 'Where is everybody?' He did some quick calculations in his head, concluding that extraterrestrial life was not only probable, the Earth should have already been and many times. So then where were the aliens? For this lunchtime insight, the Fermi Paradox was named in his honor.

A New Science of Mental Health

On the cover of the May 1950 issue of Astounding Science Fiction, there's a creature like a caveman with cat's eyes, hairy arms folded high, thick metal bracelets crossed at the wrists. He is wearing a blue fur tunic with a broad leather belt. The words 'Dianetics A new science of the mind' aim the reader's eyes directly at the space caveman's gigantic eye-shaped belt buckle.

L. Ron Hubbard's book 'Dianetics' was published the same month. A new religion was born.

The Ordeal terminates by failure - the occurence of sleep invincible - or by success, in which ultimate waking is followed by a final performance of the sexual act. The Initiate may then be allowed to sleep, or the practice may be renewed and persisted in until death ends all. The most favourable death is that occurring during the orgasm, and is called Mors Justi.

- Aleister Crowley, LIBER CDLI: Of Eroto-comatose Lucidity

Modern man has lost the option of silence. Try halting sub-vocal speech. Try to achieve even ten seconds of inner silence. You will encounter a resisting organism that forces you to talk. That organism is the word.

- The Ticket That Exploded (1962)

CREEP

In 1972, CIA director Richard Helms ordered the destruction of all documentation related to MKULTRA.

In 1972, U.S. health officials admit that African-Americans were used as guinea pigs in the Tuskegee Study of Untreated Syphilis in the Negro Male.

In 1972, the International Astronomical Union named a crater on the dark side in honor of Jack Parsons.

In 1972, the band Steely Dan chose their name based on a dildo in Naked Lunch.

In 1972, a Hungarian-born Jewish Australian geologist named Laszlo Toth attacked Michelangelo's Pietà statue with a hammer shouting, 'I am Jesus Christ, risen from the dead!'

Laszlo Toth

The alias used by Don Novello (Father Guido Sarduci) in his letters.

From: Lazlo Toth - March 13, 1974
To: Mr. Rawleigh Warner Jr., President, Mobil Oil Corporation

I would like you to know that many Americans appreciate all the oil companies have done for this country and want you to know that just because the press plays up people complaining, a lot of people know the oil crisis is not your fault any more than it is our President's. There just isn't enough oil, why can't people understand that? Don't be discouraged, the American people will someday see that you were telling the truth! God bless your people all over the globe!
Stand up for our President!
An American, Lazlo Toth

To: Lazlo Toth - February 28, 1974
From: Thomas J. Fay, Manager, Corporate Services, Mobil Oil Corporation

Dear Mr. Toth: Mr. Warner has asked me to thank you for your very gracious note of February 15. With all the criticism we have been receiving lately from some areas of the public, the press, and the government, it is nice to know that we have support from people like yourself. Thank you again for writing. Sincerely...

Our William Tell Routine

Pauline Robinson Pierce married publisher Marvin Pierce in 1919. On June 8, 1925, Marvin and Pauline had their third child, Barbara.

September 23, 1949, Marvin lost control of his car as he tried to keep a cup of hot coffee from sliding onto his wife. The car crashed into a stone wall, killing Pauline instantly.

January 6, 1945, Barbara Pierce married George Herbert Walker Bush. Their first son, George Walker Bush was born July 6, 1946.

November 6, 1963, Laura Welch ran a stop sign, broadsiding another car. She killed the other driver, who happened to be her classmate and ex-boyfriend Michael Dutton Douglas.

January 30, 1976, President Ford made George H. W. Bush the Director of the CIA.

November 5, 1977, Laura Welch married George Walker Bush

November 25, 1981, George and Laura had twin daughters, Barbara and Jenna.

May 29, 2001, Barbara was charged with attempting to use a fake ID with her paternal grandmother's maiden name (Barbara Pierce) to purchase alcohol.

May 29, 2001 - the same day his daughter was caught trying to use her grandmother's name to buy booze - George W Bush announced his refusal to stop Enron from shutting down the California power grid. 'We will not take any action that makes California's problems worse and that's why I oppose price caps.'
Around the same time, Enron traders recorded bragging about the money they stole from poor grandmothers in California. 'Yeah, now she wants her fucking money back for all the power you've charged right up, jammed right up her asshole for fucking $250 a megawatt hour.'
On May 25, 2006, Kenneth Lay was found guilty on all six counts of conspiracy and fraud by a jury, and in a separate bench trial, found guilty of four counts of fraud and false statements. He was facing up to 40 years in jail.
On July 5, 2006, Kenneth Lay 'died.' His ashes were buried in a secret location. Because he died prior to exhausting his appeals, his conviction was abated.
In 2007, three of the hundreds of Enron electricity traders who jammed their fictions up the asshole of grandma were convicted. They were each sentenced to brief probation and small fines.

Solutions to the Fermi Paradox

There are no other civilizations in our galaxy, human beings are a cosmic fluke, and that which we see before us is the very best the Universe can do. Alone. So very alone...
Or... This is what God wants, so shut up!
Or... The Prime Directive, or 'Zoo' Hypothesis. No talking to the monkeys.
Or... Other civilizations do exist and for some reason we just can't detect the evidence of the existence with our primitive monkey brains.
Or... Other civilizations do exist and the evidence is everywhere, but we don't realize it. Most likely because someone is totally fucking with us.

> *Hio, for the sunsets glowing rose*
> *Hio, for the glorious opal dawn*
> *Pueblo land of the caveman vanished long*
> *Mystic land hear the echo of our song*
> *Los Alamos!*

Jason Brown

FRAGMENTS

The following script is based on the final two pages of a letter by L Ron Hubbard, dated May 14, 1951 and addressed to 'THE ATTORNEY GENERAL.'

http://www.xenu.net/archive/FBI/fbi-110.html

Full FBI file for L Ron Hubbard:

https://secure.wikileaks.org/wiki/Scientology:_2,826_pages_of_L._Ron_Hubbard%27s_FBI_Files,_1943-1993

Laffayete Ronald Hubbard, Wichita, Kansas May 14, 1951
To: The Attorney General, Department of Justice, Washington DC

I only know these things! While I let them, unsuspected, cluster around me, these people stopped Dianetics in its tracks. With them gone, we can run an organization.

But once ejected, they began - (SNEERINGLY) evidently through Sara - these remarkable attacks.

I believe this woman to be under heavy duress.

She was born into a criminal atmosphere, her father having a criminal record. Her half sister was an inmate of an insane asylum.

She was part of a (WISTFUL) free love colony in Pasadena.

LOOK AWAY. THOUGHTFUL PAUSE...

She had attached herself to a Jack Parsons

(QUEENY) the rocket expert - during the war, and when she left him, he was a wreck. Further, through Parsons, she was 'strangely intimate' with many scientists of Los Alamos.

TOUCH SELF. NIPPLE?

I did not know or realize these things until I myself investigated the matter.

(OMINOUS) She may have a record.

PICK UP LARGE BOOK

My plea is simply this:

BRANDISH BOOK

Security! In which science can work! Why do these people remain at large?! Free of our press? Destructive of our efforts?

I have been developing - in spite of these enturbulences! - data of some value as this rudimentary pamphlet proves!

DROP BOOK

Dianetics and the Foundation - potent forces! - almost fell into complete Communist control (OFFHAND) or the control of ex-Communists, whichever it is.

I cannot fight the battle of Communist vs the World as the only opponent or threat! (PLEADING) Certainly someone else must be at least faintly interested?

My life has been in danger, my work has suffered,

CLUTCH AT THROAT

My life is still in danger.

My reputation is almost ruined so these vermin Communists (OFFHAND) Or ex-Communists, whatever they are... can take over a piece of society! And a technology!

PULL OUT NOTES

If Russia possessed the notes I have on psychological warfare, she would be that much more potent.

CRUMPLE NOTES, HIDE THEM

Further, I do not believe these people meant to destroy Dianetics but to drive it underground.

They 'helped' me with radio programs which did not get played, by pamphlets which did not give the whole story, and by 'advice' which attempted (INCREDULOUS) to knock every loyal American out of Dianetics!

What can one do in the face of this? When? When? WHEN will we have a round-up??

Please compare these notes with your central files. I am certain you will find these names repeated there connect with Communist activities.

PICK UP VOODOO DOLL

Perhaps in your criminal files or on the police blotter of Pasadena you will find Sara Elizabeth Northrup

BEGIN STABBING DOLL

Age about 26, born April 8, 19... something, about 5'9', blonde-brown hair, slender. My own investigation seems to indicate that possibility.

Her residence from '42 to '45 was (WISTFUL) 1003 South Orange Grove, Pasadena, California.

LOOK AWAY. THOUGHTFUL PAUSE...

SUDDENLY DROP VOODOO DOLL

I have no revenge motive! Nor am I trying to angle this broader than it is. I believe she is UNDER DURESS! That they have something on her! And I believe that under a grilling...

MAKE FIST, GRIND JAW

she would talk, and turn State's evidence.

PICK UP BEER AND PILLS

All these matters are, of course, confidential.

PILLS + BEER

I do not wish them to be published in any way!!

FINISH BEER, CRUSH CAN

I am not trying to regain a reputation by blaming Communism! But I am trying hard to understand...

DROP CAN

How is it that these persons, all so solidly ex-affiliated or currently affiliated, as a group work in such close partnership against a technology they know would hurt Communism (CRAZED INCREDULITY) and yet they remain at liberty??

PICK UP BRAIN

I am applying to the Department of Defense for permit to deliver them my work on psychological warfare.

POKE BRAIN

I hope this new Foundation can operate. Frankly, from what has happened,

LET BRAIN FALL

(DESPAIR) I am not certain I will live through this...

(SUPER HAMLET!) If I do not, know that I have only these enemies in the entire world!

May I respectfully request, sir, ASSISTANCE in rendering America a trifle safer for new sciences! I wish I could ask you to extend that clause in the charter of the FBI about persons in distress.

Sincerely. L. Ron Hubbard

Reesa Brown / Kit O'Conell

WHAT IS THE 21ST-CENTURY NOVEL?

Introduction: The Future of Storytelling

When one thinks of a writer, the image that often springs to mind is that of the reclusive artist, toiling alone in the dark over his typewriter. Many writers deliberately cultivate a luddite image, proudly declaring their inability to grasp modern technology. Even in a forward-thinking field like science fiction, fear of the future is rampant; this backward attitude is exemplified by former *Science Fiction Writers' of America* Vice-President Howard V. Hendrix. On April 12, 2007 he infamously declared his alliance with the past in a LiveJournal posting where he claimed that those who give their work away for free online were 'webscabs.' Opposed to the increasing use of the Internet by his fellow authors, he complained, 'I think the ongoing and increasing sublimation of the private space of consciousness into public netspace is profoundly pernicious. For that reason I don't much like to blog, wiki, chat, post, LiveJournal, or lounge in SFF.net. A problem with the whole wikicliki, sick-o-fancy, jerque-du-cercle of a networking and connection-based order is that, if you 'go along to get along' for too long, there's a danger you'll no longer remember how to go it alone when the ethics of the situation demand it.'[1]

Statements like this make writers seem poised to repeat the mistakes of the music or movie industries, resisting innovation and fearing rather than adapting to the changes technological progress inevitably brings. For the authors of this paper, Hendrix's words were the beginning of a long conversation about what other writers and artists were doing to use the new medium of the Internet to bring storytelling into the 21st-century. In the end, it led to the birth of the *Continuous Coast project*, our own attempt to use what we'd learned to create innovative Internet-based storytelling.

A Note on Sex

Many of the first uses of any new technology are inevitably to explore its potential applications toward sexuality and the erotic. Within 200 years of its invention, Gutenberg's press was being used to publish erotica. Many of the earliest novels, such as *The Tale of Genji, Canterbury Tales, Tom Jones,* and the *Decameron* intimately involve sexuality in their plots. Some of the earliest film and photography had erotic subjects, and the proliferation of pornography on the Internet is so well-documented that it is often lampooned in popular culture.[2]

Sexuality in print media has continued to flourish. At the close of the 20th-century, specialty presses such as *Cecilia Tan's Circlet Press* offer science fiction erotica to discerning readers. Supernatural romance, a genre that combines the erotic with the trappings of horror or dark fantasy, is flourishing in popularity. According to erotica writer M. Christian, the artificially-imposed boundaries between erotica and the mainstream are vanishing as erotic topics appear more and more in 'normal' genre and literary fiction.[3]

[1] Howard V Hendrix: 'Howard V. Hendrix, SFWA's Current VP'. Science Fiction and Fantasy Writers of America. 12.04.2007, http://community.livejournal.com/sfwa/10039.html

[2] For example, see Robert Lopez and Jeff Marx: 'The Internet is for Porn' in the Broadway show Avenue Q.

[3] M. Christian: 'Confessions of a Literary Streetwalker: The End of Erotica'. Imagination Is Intelligence With an Erection. 07.07.2008, http://zobop.blogspot.com/2008/07/confessions-of-literary-streetwalker.html

Looking Forward

Many of the earliest attempts at 21st-century storytelling are similar to those of the 20th-century, but translated onto the Internet. Books and magazines are published in exact digital analogs of their print form, often for about the same price. To the authors, this is analogous to the early days of television when most TV shows were essentially radio shows with video cameras pointed at them. Who, we wondered, was beginning to move further, to take advantage of the Internet as a medium deserving of new forms of storytelling?

In our research we quickly discovered that it is important to judge 21st-century storytelling by its own merits - cutting-edge projects must be evaluated based on how well they succeed at their own goals, not the goals of books, magazines or other media of the past. Cory Doctorow, one of the most forward-thinking minds in modern science fiction, notes 'the critics of new media often point to its failure to live up to the standards of old media. Some scientists and science journalists wring their hands at the idea that the Mars landers ... emanate information in the form of anthropomorphized Twitter messages, arguing that these messages lack the formal virtues of science reporting and papers. It's true. They do. They don't succeed at being better in-depth science articles than the science articles. They succeed at being better Twitter messages than science articles . . . The low cost of deploying new media online is revealing a heretofore unsuspected appetite for stories in different boxes than we've heretofore used - and a universe of stories waiting to be told.'[4]

With that in mind, we begin this paper with our survey of some of the most notable early experiments in 21st-century storytelling. As we outline each example, we will also attempt to point out a few lessons we think can be learned from each project. In the second section of the paper we'll touch briefly on the need for a business model for 21st-century storytelling. Finally, in the third section we'll discuss our own project, Continuous Coast.

Explorations of 21st-Century Storytelling

Stephen King

Many authors of conventional print fiction have begun to explore the potential of the Internet. Best-selling horror writer Stephen King was one of the earliest authors to discover some pitfalls of moving from traditional media. His novella *Riding the Bullet* was initially available exclusively in electronic form in 2000. For the first week it was totally free and the response overloaded the servers where it had been published. The book was published in a DRM (Digital Rights Management, or copy-protected, format) but this was soon broken by crackers.[5]

Emboldened by this early success, he began publishing his serial novel *The Plant* on his website later that year. He asked his readers to pay $1 per installment of the novel on the honor system; if enough readers paid, he would continue to publish the book. Over time, fewer and fewer readers paid for the novel and it remains unfinished; the last available installment was published in mid-2001.[6] Although we believe King made a wise decision in releasing The Plant in a DRM-free

[4]   Cory Doctorow: 'Don't Judge New Media by Old Rules'. Internet Evolution. 22.09.2008, http://www.internetevolution.com/author.asp?section_id=479&doc_id=164252&

[5]   'Riding the Bullet'. Wikipedia: The Free Encyclopedia. 12.10.2008, http://en.wikipedia.org/wiki/Riding_the_bullet

[6]   'The Plant'. Wikipedia: The Free Encyclopedia. 28.11.2008, http://en.wikipedia.org/wiki/The_Plant

format, we think he made the mistake of asking readers to pay multiple times for the same product. In our opinion, fans prefer to pay only once for access to a complete story, regardless of the number of installments in which it is released.

Jim Baen's Universe

Begun in 2006, *Jim Baen's Universe* is an online magazine that largely translates the print model to the web. Much like a print magazine, readers can buy individual issues or subscribe for an entire year. The magazine goes beyond the conventional offerings by giving subscribers access to previews of upcoming books from Baen Books, its publisher. We believe *Jim Baen's Universe* has succeeded in part because of its strong anti-DRM attitude. By offering extras that would be unavailable to subscribers of a print magazine, they make a subscription more valuable to flighty online consumers.

Pixel-Stained Technopeasants: Free Fiction Online

Despite Hendrix's attack on webscabs, many authors have found success through providing their fiction free online. Cory Doctorow, as previously mentioned, is among the most successful. A winner of the John W. Campbell Award for Best New Writer and a Nebula Award nominee, Doctorow has made every one of his novels available for free download via his website under Creative Commons open-content licenses. Publishers have also gotten involved: *Baen Books* created the *Baen Free Library*, a collection of free e-books available both online and as CD-Roms inserted into many of their books. *Tor Books* recently created *Tor.com*, a social networking site for their fans which the publisher uses to release free novels and stories.

Thanks to Hendrix's diatribe, many forward-looking authors gathered to celebrate *International Pixel-Stained Technopeasant Day*[7] on April 23, 2007, on which day they gave away short stories or entire novels online. Catalog websites like *Free Speculative Fiction Online*[8] provide access to thousands of stories, and every Friday the popular science-fiction blog *Futurismic* offers a round-up of the latest online offerings. Science fiction seems to be thriving online, and from this we conclude that if you give it away, people will still pay you for it.

Podcasts

Audible storytelling has seen a boom in the form of podcasts, or Internet-based audio programs. Like audiobooks before them, podcasts reach new audiences in new ways. Markets like *Escape Pod*, *Pseudopod* and *Podcastle* pay their authors a professional rate for audio reprints of their previously published print fiction. Other programs offer 'radio show'-style serial dramas, such as the popular *Metamor City* podcast. Like many online projects, podcasts earn money primarily through donations or the sales of merchandise,

The logo of the Metamor City podcast, 2008, © Metamor City Podcast

---

[7]   A name taken from the portion of his LiveJournal post which refers to 'the downward spiral that is converting the noble calling of Writer into the life of Pixel-stained Technopeasant Wretch.' The collected contributions can be found at http://duskpeterson.com/technopeasant/

[8]   http://freesfonline.de/

sometimes including CDs of past programs. Podcasts are also an excellent example of the inherently multimedia nature of 21st-century storytelling, which can no longer sit inertly on a page. Podcasts profit by cultivating a niche audience of dedicated followers, a technique sometimes referred to as the '1000 True Fans' model.[9]

Webshows and Lonelygirl15

Lonelygirl15's Bree with her monkey in one of the blog-style episodes of the show, 2007, © EQAL, Inc.

An example of an early success in video-based online story-telling was *Lonelygirl15*; beginning in June 2006 and lasting through 2008, viewers initially thought they were watching the nonfiction video diaries of a normal teenage girl. Over time, however, supernatural elements began to slip into the story and by September 2006 its fictional nature was well-known. The show quickly collected a rabid following that was so devoted to the show they began making their own fan videos. The quality of these videos was so high that some became part of the 'canon' storyline of the show. Further stretching the boundaries of traditional storytelling, the finale of the series' first story arc took place in real-time, with a new episode appearing throughout the course of the day in which they were purported to 'happen.' We believe that *Lonelygirl15* shows that 21st-century storytelling is not static but can both involve its viewers and the passage of time. This blurs the traditional boundaries between official and unofficial, or canon and 'fanon'[10] products, and involves its fans directly in the experience in ways which static storytelling does not.[11]

Webshows can also bridge the gap between traditional media and new media. Many television programs now offer abbreviated online episodes that coincide with the plotline of their normal broadcast or cable programs. *Battlestar Galactica* and *Heroes* are among the programs which have used 'webisodes' to maintain viewer interest between seasons or during writers' strikes. In some cases, a new medium is used to help transition between traditional media: this was the case with the *R. Tam Sessions*, a series of virally leaked videos linking the popular science fiction TV series *Firefly* with *Serenity*, the major motion picture which served as its conclusion; these videos helped generate buzz and prepare the series' loyal fanbase for what was to come.[12]

[9] This model proposes that a project will succeed if it meets a goal of 1,000 fans willing to each contribute $100 a year to your project. See Kevin Kelly: '1,000 True Fans'. The Technium. 04.03.2008, http://www.kk.org/thetechnium/archives/2008/03/1000_true_fans.php

[10] Fanon is defined as a continuity which exists primarily in the minds of the fans. Examples of fanon might include the lengthy explanations fans invented to explain the differences between Klingons in the classic Star Trek and Star Trek: The Next Generation universes.

[11] 'Lonelygirl15'. Wikipedia: The Free Encyclopedia. 09.02.2009, http://en.wikipedia.org/wiki/Lonelygirl15

[12] 'R. Tam sessions'. Wikipedia: The Free Encyclopedia. 13.01.2009, http://en.wikipedia.org/wiki/R._Tam_sessions

Alternate Reality Gaming (ARGs)

21st-century storytelling not only blurs the lines between canon and fanon, but between fantasy and reality itself. In Alternate Reality Games, also known as ARGs, players explore stories that unfold not just online but offline as well, through emails, phone calls, and visits to real-world locations. Due to the large budgets involved, almost all of these games have been commercial enterprises, usually as viral marketing for a future commercial project; one example was the well-known ARG *I Love Bees*, which promoted the videogame *Halo 2*.

ARGs draw their players into intricate storylines that cross many mediums and genres, challenging fans to decipher complicated puzzles. During the *Dark Knight ARG*, a list of addresses was discovered. These addresses led the players to bakeries from which the first player to arrive received a cake containing a cell phone. When called, the phone gave them a message from the Joker that led on to the next online clue.[13] *Nine Inch Nails: Year Zero* involved dozens of websites, fliers and graffiti, as well as USB thumb drives left at *Nine Inch Nails* concerts which contained specially altered songs. At the end of the game, a few lucky fans got to visit a strange warehouse where they were treated to a special concert by Trent Reznor, only to have the concert broken up by a fictional SWAT team.[14] 21st-century storytelling clearly asks its fans to experience story in ways more challenging than simply turning the pages of a printed book.

A poster from the the Nine Inch Nails: Year Zero Alternate Reality Game, photographed in the wild by a user on flickr, 2007, photo © Nic Walker, flickr user nic0

Dr. Horrible's Sing-Along Blog

Dr. Horrible's Sing-Along Blog is perhaps the first commercial online musical; it was first conceived by Joss Whedon (best known for TV shows *Buffy the Vampire Slayer*, *Firefly*, *Dollhouse*, etc.) with friends and relatives during the 2007 *Writers Guild of America* strike. Originally offered as three 12-minute episodes which were available for viewing over the course of a single week, the commercial download of the series became a best-seller on iTunes even while the episodes were available for free. To further encourage fans to pay for this free product, various tie-ins such as t-shirts were offered, almost all of which appear in the videos themselves. Fans could also become more involved with the world by applying to become supervillains in the *Evil League of Evil*, further encouraging them to follow the series and its products. Once again, Dr. Horrible proves that for 21st-century storytellers, freely products can be quite profitable for their creators.

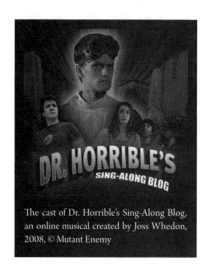

The cast of Dr. Horrible's Sing-Along Blog, an online musical created by Joss Whedon, 2008, © Mutant Enemy

---

[13] Maureen McHugh (Consultant for 42 Entertainment, producers of many ARGs), in discussion with the authors, 2008.

[14] 'Year 0 Case Study'. 42 Entertainment. 26.07.2008, http://www.42entertainment.com/yearzero/ and 'Year 0 (game)'. Wikipedia: the Free Encyclopedia. 12.01.2008, http://en.wikipedia.org/wiki/Year_Zero_(game)

Shadow Unit and Alternate Reality Fiction

Applying the lessons of Alternate Reality Gaming to more traditional forms of fiction is a logical next step. In 2007, successful authors Emma Bull, Will Shetterly, Sarah Monette, and Elizabeth Bear, along with artist Amanda Downum,

The logo of Shadow Unit, fanfiction for an imaginary TV show, 2007, © Amanda Downum

launched *Shadow Unit*. *Shadow Unit* takes the form of 'fanfiction' for a fictional TV show that is part *X-Files* and part *Criminal Minds*. During its first 'season', Shadow Unit's creators offered weekly fiction; in the season finale, a novel was released in parts over the course of a week, with each part taking place on the day it was released.

Like many 21st-century storytelling projects, *Shadow Unit* intensifies fan involvement by drawing them into its fictional world. Major characters from the series keep LiveJournals where fans can interact with these fictals[15] as if they were real people. One fan, TxAnne, even briefly became the online girlfriend of Chaz Villette, one of the stars of the show. Fanfiction is encouraged under the series' Creative Commons license. A liminal space for experiencing the series is further created through 'Easter eggs' hidden throughout the website, including a photocopied 'shooting script' from the TV show, complete with the imaginary actor's marginal notations.

With one of five projected seasons complete, and the next scheduled to begin in 2009, Shadow Unit seems an early success in Alternate Reality Fiction (ARF).[16] Shadow Unit also illustrates some of the unexpected challenges of new forms of storytelling. When the fictal Chaz Villette's LiveJournal vanished temporarily due to on-going events in the plot of the series, fans reacted with the same horror that accompanies the deletion of a nonfictional friend's LiveJournal. Some felt that a real friend had left their lives or that their enjoyment of Shadow Unit had been impaired, while others simply lamented the loss of access to all of Chaz's recipes. What was planned as a normal result of character development had unforeseen reactions among the fanbase which could not have been predicted based on old forms of storytelling.[17]

Other ARF projects have begun to appear, including our own *Continuous Coast project* which is detailed later in this paper. Another recent arrival is *Runes of Gallidon*, a high fantasy world open to any creators; *Runes of Gallidon* encourages fan creators to tithe a portion of their profits back to the project in return for creating in their world as part of its 'Artisan Agreement'.[18] As time goes on, the authors of this paper expect more worlds to appear online, open for contributions from both creators and fans, and further blurring the distinctions between the two.

[15] A word coined in 2008 by Reesa Brown, a fictal is a fictional character that one can interact with online in environments such as blogs, social networks, or forums. See Reesa Brown: 'CC Fiction meta-discussion'. The Continuous Coast project beta forums. 02.11.2008, http://continuouscoast.com/forums/viewtopic.php?f=4&t=32&p=486

[16] A term coined by Andrew Plotkin to refer to Shadow Unit, the Continuous Coast Project, and other similar projects. See Andrew Plotkin: 'Alternate reality fiction'. The Gameshelf. 08.10.2008, http://gameshelf.jmac.org/2008/10/alternate-reality-fiction.html

[17] 'What happened to Chaz's LJ?' Shadow Unit forums. 23.10.2008, http://www.shadowunit.org/smf/index.php?topic=478.45

[18] 'Artisan Agreement'. Runes of Gallidon. 2008, http://runesofgallidon.com/artisanagreement

Where's the Sex?

So far, our exploration of 21st-century storytelling has uncovered precious little sexuality, a further sign of the medium's immaturity. Usenet newsgroups and, more recently, websites such as literotica.com or the collaborative storytelling available on writing.com, have long offered a venue for sharing erotic stories but there seem to be few projects pushing the boundaries of online storytelling. Exceptions are beginning to appear, such as *The Fold*, an online webshow created by erotica authors Polly Frost and Ray Sawhill in collaboration with filmmaker Matt Lambert. A sci-fi epic told in six episodes, the webshow brings a burlesque-like raunchiness to online storytelling. Even with exceptions like this, we believe online storytelling may face challenges when it comes to sexual maturity, not unlike those which have been faced in the comic book industry[19] or video games.[20]

Towards an Online Business Model

One issue hampering the development of 21st-century storytelling is the lack of a clear business model through which artists can profit from their creations online. *Shadow Unit* has become a full-time job for its creators, which means that 5 people must make their living largely off of donations that averaged about $1000 per month during season one.[21] When the authors of this paper attended *Fourth Street Fantasy Convention* in June 2008, we found numerous industry professionals all asking the same question: how do we make money online? The old models are dying and the new ones are as yet unproven.

A scene from episode 6 of The Fold, a web series created by Polly Frost, Matt Lambert, and Ray Sawhill, © Rapture House and Mallard Media

Based on our research, we believe that there is not one single answer, but that instead online creators must tailor a business model from the overall toolkit of available income streams. While creators should accept donations, it is important to diversify into other ways of encouraging fans to give back to their creations. It is important to involve your fans by making them feel as if their contributions not only support the creators but involve them in another world itself, whether by allowing the fans to own something they saw in the work itself (such as the t-shirts of *Dr. Horrible's Sing-Along Blog*) or by turning the fans themselves into creators. Some of our own speculations have been published online on our blog.[22]

[19]   For one example of this issue as seen in comics, see Gail Simone: 'Women in Refridgerators'. 03.1999, http://www.unheardtaunts.com/wir/

[20]   For an examination of sexual maturity in video games, see Daniel Floyd: 'Chasing Maturity: Video Games and Sex'. Boinkology. 28.05.2008: http://boinkology.com/2008/05/28/making-adult-video-games-a-little-more-adult/

[21]   Emma Bull and Will Shetterly (creators of Shadow Unit) in conversation with the authors, June 2008.

[22]   See Reesa Brown and Kit O'Connell: '21st-Century Artist Business Models'. Words Words Words. 2008, http://dreamcafe.com/words/category/artist-business-models/

The Continuous Coast Project

After spending so much time studying and talking about 21st-century storytelling, it was only natural that we'd want to try it for ourselves. We asked ourselves how far we could push the models of online fiction we'd uncovered in our research. We knew that any project we created had to reflect the themes, ethics, and storytelling paradigms which we wanted to explore, disrupt, and exploit - both in the structure of the project as well as the stories themselves. As a result of this brainstorming,

Austin, Texas-area musician Gfire performs at Lufton Runner, a concert venue from the Continuous Coast, 2008, © Amul Kumar and Gfire

we began the prologue (or beta test) of the *Continuous Coast project* in late 2008. The project started in collaboration with our housemate, Steven Brust[23], but we quickly realized it was bigger than just the three of us. The project now involves many other writers, artists, artisans, computer geeks and other creators.

Shared worlds have been explored before in traditional print storytelling. *Thieves' World*, created by Robert Lynn Asprin in 1970 involved numerous writers in over a dozen anthologies and novels collectively exploring a city at the edge of a fictional empire.[24] More commercial properties such as the *Star Trek* or *Marvel Comics* universe have been explored in novels and short stories as shared worlds; indeed, the common superhero universes of *DC* and *Marvel Comics* are shared worlds with a common cast of characters that have become familiar to readers over the course of decades.[25] Alternate history author *Eric Flint's 1632 Universe* has become a collaborative enterprise in recent years, with approved fan contributions being published in the online magazine, the *Grantville Gazette*. However, all these worlds are released under a standard closed copyright, maintaining the artificial distinction between the official creations of the copyright holders and the unofficial and often literally illegal creations of its devoted fans.[26]

Even major editors have decried this ridiculous legalistic distinction that makes certain works unpublishable. Teresa Nielsen Hayden, long-time editor at *Tor Books*, wrote: 'In a purely literary sense, fanfic doesn't exist. There is only fiction. Fanfic is a legal category created by the modern system of trademarks and copyrights. Putting that label on a work of fiction says nothing about its quality, its creativity, or the intent of the writer who created it. . . . I'm just a tad cynical about authors who rage against fanfic. Their own work may be original to them, but even if their writing is so outré that it's barely readable, they'll still be using tropes and techniques and conventions they picked up from other writers. We have a system that counts some borrowings as legitimate, others as illegitimate. . . . Personally, I'm convinced that the legends of the Holy Grail are fanfic about the Eucharist.'[27] Our own collaborator and best-selling author of the expansive Dragaeran series, Steven Brust, wrote, 'I had an epiphany: when I write a second novel set in a world I created, I'm writing fanfic. . . . fanfic is fiction written by someone who is geeked by the original creation and wants to continue it. That'd be me.'[28]

[23] http://www.dreamcafe.com/

[24] See 'Robert Lynn Asprin'. The Encyclopedia of Fantasy. New York: St. Martin's Press 1997, p. 65 and 'Thieves' World'. Wikipedia: The Free Encyclopedia. 21.12.2008, http://en.wikipedia.org/wiki/Thieves%27_World

[25] See 'Shared Worlds'. The Encyclopedia of Fantasy. New York: St. Martin's Press 1997, p. 859.

[26] 'Shared world'. Wikipedia: The Free Encyclopedia. 17.01.2009, http://en.wikipedia.org/wiki/Shared_world

[27] Teresa Nielsen Hayden: 'Fanfic: force of nature'. Making Light. 25.04.2006, http://nielsenhayden.com/makinglight/archives/007464.html

[28] Steven Brust: 'More scattered thoughts on Fourth Street.' Words Words Words. 25.06.2008, http://dreamcafe.com/words/2008/06/25/more-scattered-thoughts-on-fourth-street/

Open Content

So why not create a shared world that attempts to remove these barriers completely? For the *Continuous Coast project*, we selected the *Creative Commons Attribution-Sharealike 3.0 license*.[29] This license allows for others' to make full commercial reuse of all aspects of our project, including 'remixing' or the creation of derivative works, as long as those derivative works are released under a similar sharealike license. While many 21st-century storytelling projects make use of a noncommercial license - for example, Cory Doctorow actively encourages noncommercial remixing of his novel *Little Brother* - it is unusual to allow full commercial reuse.

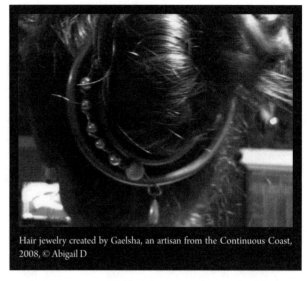

Hair jewelry created by Gaelsha, an artisan from the Continuous Coast, 2008, © Abigail D

We believe that the best stories of the 21st-century will be inherently collaborative in nature, involving not just a single writer, but a network of writers, artists, and other creative people working on concert. The open and viral nature of the sharealike license not only encourages others to join in the act of creation, but means that their works will also further inspire, creating a pool of works related to the *Continuous Coast* which can be drawn on at any time.

The desire to control reproduction of content is an antiquated one in our modern digital world. Instead, we must look for new ways to profit from our creations. In 'Better than Free', Kevin Kelly writes that the Internet is a 'copy machine' and we must look for things that can be sold which are not able to be copied. Fans will pay for something which is authentic, and which makes them feel involved as more than just spectators (for an example of which we can again look back to *Dr. Horrible* and *Shadow Unit*). We must find ways to encourage fans to easily give back in ways that satisfy both the creator and the viewer and which stretch beyond simply downloading a book in .pdf format.[30]

Blurring the Lines

In Jorge Luis Borges' story, 'Tlön, Uqbar, Orbis Tertius', the protagonist (a fictal version of the author) details his discovery of a grand conspiracy to create a fictional world that will eventually replace our own. Over time, the artifacts and places of the world begin to cross over from fiction into reality, first in *Encyclopedia Britannica* entries, then later as real physical objects.[31] In the *Continuous Coast project*, our fans have the opportunity to not only read static fiction from the world, but also to interact with the personalities of many of its residents. They can view photographs of real musi-

A bowl for the consumption of food created by native ceramic artist Debs and imported from the Continuous Coast, 2008, © Melded Earth

[29] http://creativecommons.org/licenses/by-sa/3.0/us/

[30] Kevin Kelly: 'Better than Free'. The Technium. 31.01.2008, http://www.kk.org/thetechnium/archives/2008/01/better_than_fre.php

[31] As of this writing, a side-by-side Spanish with English version of this story is available at http://interglacial.com/~sburke/pub/Borges_-_Tlon,_Uqbar,_Orbis_Tertius.html

cians playing concerts at our fictional concert venue, Lufton Runner.[32] Rather than offer simple branded merchandise, we instead offer actual artifacts imported from the *Continuous Coast* - already artisans have created dinnerware, scarves, and jewelry for export onto Earth. And these artists in turn inspire the writers - when *Melded Earth*, our ceramic artist, presented us with a set of plates, bowls, and small drinking cups from *Continuous Coast*, we immediately discovered that the citizens of our imaginary world eat in a communal style, with small portions shared 'family-style' from large collective platters and pitchers. By inviting other artists into the formerly solitary act of creation, the result is a richer world.

In the future, we will also offer citizenship in one of our fictional communities, Port Outreach, complete with 'citizenship numbers'; this will encourage our fans to question to which world they belong. Readers will be offered the chance to fund expeditions into the unexplored portions of the world, in return for which they will receive exclusive artifacts, images, and reports from the journey. In time, complete setting notes will be available publically so that anyone who wants to can add their works to the project. Although the core creators will still determine which works they deem 'canon' or 'fanon', under the terms of the *Creative Commons* license, these fan products will still be both legally marketable and a part of the collective library from which the project can draw.

World Building

Our goal in creating *Continuous Coast* was to invent a realistic world that we would want to live in, and which simultaneously reflected the ethics of the project itself and presented real challenges that writers and other creators could explore. The *Continuous Coast* is an inherently sex-positive world, both reflecting the values of its creators and encouraging the

The Mesh Scarf, created by Freya Paxtwist of Camp Peaceful Dreams, a resident of the Continuous Coast, 2008, © Andrea O'Sullivan

mature themes that have historically hastened the development of a medium. We created a setting which was post-scarcity and deliberately disposed of many standard tropes we and our collaborators had grown tired of - for example, on *Continuous Coast* there is no acceptance of rape and racism and sexism are culturally unacceptable; changes like these force our creators to look further and explore new forms of human culture while they explore 21st-century storytelling.

By challenging creators to explore new forms of conflict based in our idea of a healthier society, we introduce these ideas to our readers as well. By creating a world we'd want to live in, we also create an equally inviting shared space for those that join our project as fans. By infecting Earth with the physical artifacts of another world, we also infect it with that world's memetic artifacts. For example, when Dale Pressman, a fictal publisher from Port Outreach wrote about his unauthorized publication Emma Bull's very real novel *War for the Oaks*, a heated discussion erupted between other citizens of our imaginary world and its nonfictional fans. In order to be able to explain the customs and law of planet Earth to visitors from afar, our readers were forced to first examine these very customs and the assumptions they bring.[33] Imported scarves or fictionally pirated novels begin a conversation, and we believe inspiring challenging conversations is part of what art does.

[32] http://luftonruner.com/

[33] 'Review: War for the Oaks by Emma Bull'. Voices from Port Outreach. 09.11.2008, http://portoutreach.com/voices/2008/11/09/review-war-for-the-oaks-by-emma-bull/

Conclusion: 21st-Century Storytelling

We believe that the stereotypical (but often all-too-real) luddite science fiction writer will be unable to survive in tomorrow's online writing world. 21st-century storytelling is inherently multimedia, multigenre, and multidisciplinary. It requires a massive toolbox of creative abilities and technologies, and challenges creators and viewers to interact with art in new ways. The 21st-century novel is of course not a novel at all, but a complex network of interwoven forms, from traditional printed works to audio dramas to complete three-dimensional interactive Second Life-style environments. We do not believe the novel will die anytime soon, but it may become just one part of a reader's overall experience. Just as online media awaken 'unsuspected appetites,' they also spawn unexpected forms of storytelling, mixing the traditional and the innovative with the needs of a new century.

Stencil art appears on the door of a ship house in Port Outreach, one of the major cities of Continuous Coast, 2009, © Mary Dell

Other References

'Alternate Reality Game'. Wikipedia: The Free Encyclopedia. 02.02.2009, http://en.wikipedia.org/wiki/Alternate_reality_game

Warren Ellis: 'The Guts of Dr. Horrible'. 25.07.2008, http://www.warrenellis.com/?p=6206

Heidi Miller: 'Case Studies: Blogs' (via Amul Kumar on private Continuous Coast project forum, 2008).

Heidi Miller: 'Case Studies: Podcasting' (via Amul Kumar on private Continuous Coast project forum, 2008).

'What is the 21st-Century Novel?' Online

This paper is available online under a Creative Commons license in several formats, including Microsoft Word, as Powerpoint slides, and in audio format at http://continuouslabs.com/ae2008/

Various links from this paper can be found at http://delicious.com/todfox/ae2

Reesa Brown and Kit O'Connell continue to give this paper as a presentation at conventions and conferences. We are interested in hearing your comments or about new forms of storytelling you've discovered online. You can contact Reesa at reesa@reesabrown.com, and Kit at kitoconnell@pobox.com.

Karin Harrasser

PROSTHETICS AND FUTURE FETISHISM

Sexologist concepts of fetishism

If we think of sexuality in the sense Michel Foucault (Foucault 1988) has taught us to think about it: sexuality as the most culturally regulated, heaviest scientifically studied realm of human behaviour that is 'naturalized' in the service of biopolitics; if we think of sexuality in this framework, it is obvious that technologically enhanced sex is nothing but one variety in the whole range of 'artificial' modes of having sex. But since the beginning of sexology – which is usually associated with the name of Viennese criminal psychiatrist Richard von Krafft-Ebing – genital, reproductive sex has been considered not only the 'natural' and 'normal', but the only 'real' mode of sex, whereas all other kinds of sexuality were considered more or less 'perverse'. Consequently, cases of fetishism (with fur, hair, leather, shoes etc.) held a big share in Krafft-Ebings founding publication *Psychopathologia Sexualis* from 1886. Fetishism figured – besides homosexuality – as a 'model-perversion' for the emerging science of sexuality. Krafft-Ebings theory of the fetish was strongly influenced by literary works such as the Austrian writer Leopold von Sacher-Masoch's *Venus in Furs* from 1870 and by anthropological concepts of the fetish, which ascribed fetishism to 'primitive' cultures as a kind of surrogate for 'real', meaning monotheistic, belief.

Krafft-Ebing therefore conceptualized fetishism as surrogate-sex for 'real', reproductive sexuality. His fetishists – all of them male – were nervous characters that were not able to make the right use of their genitals but instead got enjoyment out of things 'dead'. His concept was later picked up by Sigmund Freud who conceptualized the fetish as a quite literal surrogate for the mother's missing penis. In his view, women would be ‚natural fetishists' and therefore lifetime seekers for a replacement for their missing phallus, whereas fetishist men would be perverse in the sense of a wrong identification with their mother and her missing penis. Female fetishism was therefore associated with women's general desire to decorate and robe, which was interpreted as an attempt to hide their 'defect' by showing off. It is easy to hear Karl Marx' concept of commodity as fetish, as expressed in the Capital in 1867, resonate with this concept. Male fetishism on the other hand surfaced as an illegitimate affection for a partial object, as an inadequate surrogate for the 'full' love of a woman with the clear telos of producing children to maintain the national economy.

Read with Foucault, it is more than obvious that so called 'fetishist practices' recorded by Krafft-Ebing and Freud were no expressions of an individual pathology but precise answers to the restrictive construction of heteronormative, genital sex as the only accepted mode in the late 19th century and a response to the increasing amount of commodities that leaked into peoples lives. We can then consider fetishist practices in general to be a rather emancipatory plea for the idea that *all* sex is artificial and partial; as an argument for the generative potential of perverse sex and for an understanding of sexual encounters as a meeting place for humans and non-humans. Fetishism places sexuality on the intersection of human rootedness in a biological 'wetwear' (the desire to be touched, not to be alone) and the cultural: individual and collective histories. Over time fetishism plays with acquired, culturally encoded and embodied imaginaries. But it also releases the potential to project ourselves into the future; it releases our inventiveness and our eminent non-natural side: technologies not only moderate or express our relationship to each other, but they have also profoundly altered our relationships. We can therefore think of fetishes as magical tools that can reshape our experiences. Plus the good news with fetishism is still: *we don't have to reproduce ourselves!* We are free to make use of our body and our sexuality for pure pleasure, we don't even have to use it on other humans, and we can produce things that don't resemble humans at all.

Prosthetic fetishism

How do prosthetic technologies relate to this concept of fetishism? Matthew Barney's *Cremaster 3* from 2002 will serve as a blueprint for my argument. The film is a highly condensed argument on the topics I want to discuss here. The five part

Cremaster-film-sculpture-performance-series (see Barney 2003) receives its name from the *cremaster*-muscle, whose function it is to raise and lower the scrotum in order to regulate the temperature of the testis and promote spermatogenesis. Its development is also considered to be central to the prenatal, anatomical differentiation of the sexes. In his five films Barney's aesthetic universe evolves in quite a 'bionic' way: the biological starting point – the *cremaster* muscle – is being inserted into narrative frameworks from other realms, such as biography, mythology, and geology. The films are full of anatomical allusions to the position of male and female reproductive organs during the embryonic process of sexual differentiation: *Cremaster 1* represents the most undifferentiated state, *Cremaster 5* the most differentiated. *Cremaster 3* – being the pivot of the cycle – presents images of conflicting teleologies of differentiation: images of hyperfemininity and hypermasculinity but also hermaphroditic images. And the film is full of prosthetic devices, of mythological figures and animals, which have been used for centuries to sound out the borders of humanness.

Prostheses are in all of Barney's work and are used extensively on a figurative level, being featured by certain semi-fictional characters Matthew Barney stages. For example the legendary American Football-player Jim Otto who, after numerous surgeries, ended up with two artificial knees and finally a left leg amputation in 2007. But medical-technical materials are also used as working material. Matthew Barney uses prosthetic plastics, as well as Teflon and stainless steel, for prosthetic joints in his sculptural work. Furthermore the aesthetic principles of 'restriction' and 'supplement' are central to his form finding process. His performances that take place under the label 'drawing restraint' for instance feature a restricted semantic vocabulary that is borrowed from the fields of sports (fitness, climbing, rugby), sexual perversion and mythology. Bondage scenes (inspired by free climbing and the 19th century escapologist Harry Houdini) blend with performances of exaggerated masculinity and move on to cross-dressing scenes. Change of gender is only one of the many metamorphoses that take place constantly: characters move from human to animal to mythological figure and back again. But they don't change easily: the changes are forced and often connected to moments of violence, of conflict, of rivalry.

Aimee Mullins as Entered Novitiate

The woman with the plexiglass-prostheses (who appears in *Cremaster 3* in at least five roles) is Aimee Mullins, a double amputee and athlete who has set Paralympic records in the 100 meter dash and in long jumping. She has worked as a model and actress and is very active in promoting rights for the disabled in the USA. In this scene, she plays a prosthetically enhanced Cinderella. The scene is the axis of *Cremaster 3* (and therefore the axis of the whole cycle): the chorus-line-like middle piece of the film is staged on five levels of the Guggenheim Museum and is called 'The Order'. Barney's encounter with Mullins takes place on the third level and is meant to represent the narcissistic stage of aesthetic production: Mullins serves as a kind of distorted mirror of the artist, who also wears glass shoes and is dressed in female vesture.

A minute later Mullins will transform into a hybrid of cheetah and human. This role refers to the name of her running prostheses: they are called *cheetahs* and are of the same kind Oscar Pistorius uses and that caused the International Olympic Committee so much trouble in 2008. They were classified as technological doping, Pistorius did not accept the sentence, and was admitted but by then it was too late to qualify for the regular Olympics.

Another scene within 'The Order' shows women's bodies formatted into a disciplined chorus, dressed up as sheep – no electric sheep though. A scene in which as Siegfried Kracauer once put it, the machine-logic of the military and of the factory is rearranged as a joyful 'ornament of the masses' (Kracauer 1997). In 'The order' modernity's program of aesthetic and technological alterings of the human body is exhibited in all its ambivalence: the exotic fascination with aberrant bodies, the scientific gaze onto these bodies and the proximity of a normative body image and the desire to overcome the limits of the body are condensed in highly pervasive images. The desire to alter and enhance the body via training and/or technological devices to make his owner 'fit' for ascent and advancement is allegorized in Barney's climbing-exercises within the Guggenheim museum and echoed in his ascent within Chrysler Building's elevator funnel. This fascination with body alterings blends with explicit images out of the archive of sexual perversion: bestiality, anal sadism, the reification of women bodies as pets, object-love, and/or traditional fetishes: shoes, fur, silk.

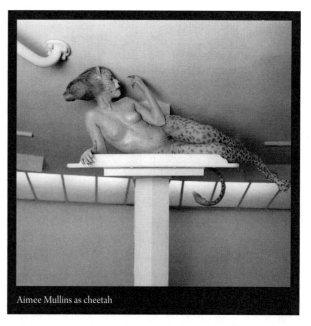

Aimee Mullins as cheetah

In this sequence prosthetic fetishism is quite obviously entangled with a rather male version of objectifying, taking apart and reassembling the human body.

But *Cremaster 3* also discusses the prosthetic altering of a male body in quite an interesting way.

In this scene Barney is punished by the Masonry; he has a strangely shaped, half amputated, half artificial penis and is placed on a dentists/gynecologists chair, where he is tortured with braces made from scrap metal taken from artfully crashed cars. But the braces are not only torture instruments and bodily restraints: they enable him to deliver teeth through the tube that extends from his anus.

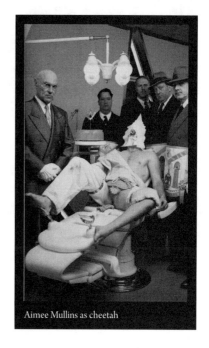

In this sequence the close relationship between virility and violence, between potency and potentiality, between perversion and creativity (making a wrong use of something), between bodily restriction and creative production is articulated quite straight away.

What Barney does here twists the sexologist's verdict of the fetishist being 'unproductive' for the societal body (because he is not engaged in proper reproduction) into a form-finding program of *artistic* production: by assembling heterogenious and highly idiosyncratic objects – mythological, technological, biological and biographical – the artwork is being deliberately developed without ever gaining a definitive form. Prostheses play a central role: they exhibit the artificiality of all form-finding-labour conducted by the different characters. But still, it's only men

Aimee Mullins as cheetah

that create art here: Richard Serra as the maitre/architect with his building of phallus-like towers and the entered apprentice Matthew Barney who challenges his hubris by experimenting with more 'feminine' form finding principles.

The dentist – the universal cripple

I would like to draw your attention to the stage of this scene, the dentist's office, to shed light on the context in which prosthetics as high-tech artifacts were developed in the first place. And I would like to hark back a bit into the history of prosthetics to show how precise this stage is used in *Cremaster 3*.

Industrial production of prosthetics started in the USA just after the civil war. Before that time prosthetics were individually crafted by blacksmiths, carpenters or manufacturers of instruments. In the 1870s new methods of mass production were developed to respond to the massive need for prosthetics as caused by the war. This 'need' was fueled by fears of losing male labour power which was connected with the fear of destabilization of gender-roles. The prosthetic industry clearly followed a patriotic goal to reassemble the social, a gesture that was literally performed by reassembling the soldier's body with prosthetics to put the maimed soldiers back into the labor market and into their families as breadwinners (Herschbach 1997).

New interfaces for workers and machines, Prüfstelle für Ersatzglieder Berlin 1919

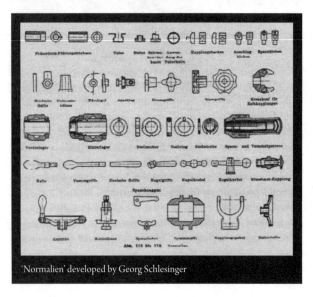

'Normalien' developed by Georg Schlesinger

The next technological advancement in the prosthetic industry took place during WWI, as the production on prosthetics were enhanced by principles of Taylorism, the European 'Science of Work' (Arbeitswissenschaft) and advanced engineering as well as by the findings of psychophysics and of so called 'psychotechnics', an early form of applied psychology (see Perry 2002, Price 1996). One of the leading engineers of that time, the German Georg Schlesinger, who was at the same time factory manager of a gun plant near Berlin, developed a new principle of interconnection with all parts of the prosthesis and the human body. By modularizing and standardizing the interfaces, he wanted to achieve the perfect ‚fit' of the prosthesis with the worker and of the worker with his machine.

Schlesinger's ideas of friction-free connections to maimed soldiers with machines were not really successful though and prosthetic technology took a different way. But what was very successful was Schlesinger's powerful manufacturing principle: the standardization of connections with so called 'Normalien' (normalcy).

His system for the replacement of organs with industrially produced, interchangeable spare-parts revolutionized prosthetics: it made the mass production of modular prostheses possible, which has since then been applied worldwide.

Staying within the early days of prosthetics of WWI, I want to return to the US-context and draw your attention to a different model of cripple-care. The famous research-couple Frank Bunker and Lillian Gilbreth were well known for promoting the Frederick Winslow Taylor system of scientific management with their film-based motion studies. The Gilbreths filmed test persons within rasterized spaces to precisely record their movements which were in a second step dissected into work steps they called 'therbligs' (an anagram of Gilbreth). These codified basic movements would then be transmitted onto charts, to identify the most efficient way to do something – for example folding handkerchiefs or packing soap.

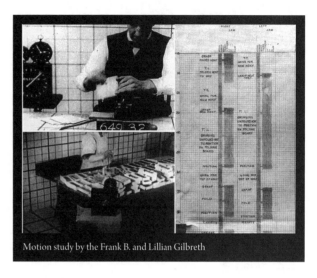

Motion study by the Frank B. and Lillian Gilbreth

Gilbreth, a bricklayer, contractor and scientific autodidact, had spent the first years of WWI in Germany to help the German army optimize procedures of war surgery. He then returned to the US in 1916/17 and together with his wife Lillian became one of the leading voices in cripple care. They developed a rather avant-garde view on what it meant to be a cripple: 'When we come to consider the subject closely we see that every one of us is in some way a cripple. (...) We can, then, think of every member of the community as having been a cripple, as being a cripple, or as a potential cripple (cit. after Brown 2002).' The consequence of their concept of the universal cripple was that they did not so much put effort into the *prosthetic repair* of the maimed bodies but to set up adaptable *working environments* that allowed all the potential cripples to perform their jobs efficiently. One of the privileged professions, one in which a technified environment would be especially suitable for adaptations in favor of the cripples was in the Gilbreth's view the dental hygienist. On this image you can therefore see a simulated cripple (see his hand stowed away under the white coat) engaged in cleaning the teeth of a patient for demonstrational reasons.

I want to interrupt my little historical digression here and return to the prosthetic fetish as presented in Matthew Barney's art work: it is almost uncanny how precise Barney's arrangement is in this case: Taylor's and Gilbreth's systems were in the beginning of the 20th century subject to harsh criticism by trade unions. Labour leaders feared – with good cause – the devaluation of skilled labour by the Taylor system and therefore strongly disapproved of the

Gilbreth's experiments with working environments for dentists

new scientific methods. They realized that the Taylor system would privilege the managers and planners over the workers. *Cremaster 3* is not only set in the 1920s, the time of the labour-fights, it is also about the Irish trade unions; and this is why the white-collar-workers torture the craftsman Matthew Barney.

Another influential post WWI image resonates all too strangely with *Cremaster 3*: it is an image the anarchist Ernst Friedrich used in his popular anti-war pamphlet *War to the War* from 1924. On the right you can (again) see Matthew Barney's character of the Entered Apprentice. The shocking photographs of maimed soldiers were in the 1920s used as visual

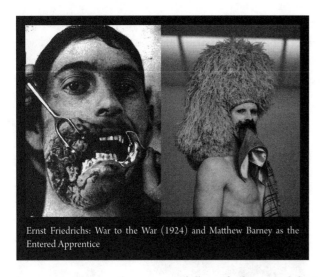

Ernst Friedrichs: War to the War (1924) and Matthew Barney as the Entered Apprentice

arguments against the devastations of the technified war. Facial prostheses were later produced to hide the wounds the war had caused to individuals and to the Prussian state. For Barney there doesn't seem to be a cure for injury, only a provisional silk drapery hides and shows at the same time the open wound.

By recounting the historical roots of prosthetics we can now identify the ingredients of Barney's poetic film-machine more precisely: it is a military-industrial-research-complex that is echoed in these sexually loaded, fetishistic images; but this complex stripped bare from its promises of a technologically and managerially rationalized bright future. Barney's prosthetic fetishes are either phallic torture instruments used by men on men or they are carefully crafted, magical devices that allow transgressions between the animal and the human, man and woman, past and future. Instead of following the functional trail of prosthetics as tools for reassembling the social and to secure production and reproduction, Barney installs a polymorphous perverse (the term was used by Sigmund Freud for the unguided pleasures of babies) universe of *artistic* production, that twists the normative notion of fetish as an artificial surrogate for 'the real sex' into a highly elaborate program of artificial restraint as a precondition for creative production.

Technofetishism

Now, what does all that have to do with technofetishism, which you might have expected to take place on spaceships and alien planets rather then in the Guggenheim museum? But I would like to argue that the Guggenheim museum is itself – like contemporary art as such – a kind of spaceship. A space set aside from the normality of everyday life, a space where imagination and experiment find a rescue. And just remember how well the Guggenheim fitted as a futuristic backdrop to *Men in Black* (Barry Sonnenfeld 1997). Science fiction's fascination with technology is thoroughly linked to military research which was as shown above the birthplace of modern prosthetics. The first notion of a Cyborg for example did not come up in a science fiction text or in Donna Haraway's famous manifesto (from 1991) but in government sponsored research on perceptional and physiological effects of outer space on humans in the 1960s (Clynes and Kline 1960). The whole idea of the necessity to enhance human features technologically seems to be deeply rooted in the figure of the armed soldier and his technological dominance over the enemy. Still, I think that science fiction – as does contemporary art – does more then just retell imperialist stories of male domination. Just think of the ambivalent image of body-enhancement that is drawn in the *X-Men* comics and films. The *X-Men* (which also comprise of women) are a bunch of mutants whose genetical and morphological otherness is subject to anxiousness in the rest of society. It is not clear whether their mutations are a gift, an enhancement or a disability that has to be protected. This leads to two groups of mutants: those who want to blend into society and use violence only for defense reasons and those (Magnetos people) who consider themselves superior to unaltered unions and therefore want to take over power. For the subject of prosthetics the character of *Wolverine* is especially interesting: he is a mutant – he heals especially fast – but has also been subject to involuntary experimental surgery that has left him with prosthetic devices, such as his infamous iron-claws. This is why he embodies the conflict over technified bodily difference and its complicated relation to the spontaneity of desires especially well.

Technology in science fiction is usually overtly sexualized (phallic forms, leather costumes etc.), although rarely ever used for experimental sexual intercourse which throws light on how technofetishism as a cultural mechanism works. It is a metaphor for a hidden knowledge about the enmeshment of sexual desire for the alien 'other', technology and domination; a knowledge that is usually not allowed to surface because it lies at the heart of power structures that combine cultural, economic and technological superiority and that culminate in the demonstration of military power.

This knowledge surfaces in Marie Chouinard's dance piece *Body remix / Les variations Goldberg* (2008). Her choreography painstakingly examines the relationship of prosthetics as extensions of the human body, their limiting and self-restricting effects as well as their productive and liberating moments. It shows dancers equipped with crutches and other orthotics tied in corset-like costumes that perform with and against each other. It is a piece with a dialectical mode: it performs the idea of the potentiality of artificially extended and willfully limited bodies without forgetting that techniques – be they external to the body or internalized like dance training – are always drawn between pleasure and violence. The desire for a different body and the limitations of what we have never line up easily and they are heavily dependent on culturally informed images and ideals. The choreography demonstrates very convincingly that we can never subtract power structures – such as gendered ones – from technofetishism. All we can do is work with and through these ambivalences.

Unlike most science fiction-films Chouinard's interest in altering and mastering the features of the 'human motor' (Rabinbach 1990) is not the creation of superhero-like 'enhancements' of the body with technological gadgets. Sometimes, the bodies here *do* gain totally new features – they are for example levitated, which is of course the old dream of ballet – but sometimes they have to fight against the fortitude and agency of the crutches and other orthotics in use. And it is not clear who the winner of this fight is, when in the end the apparatuses as well as a marionette-like dancer ascend into theater heaven.

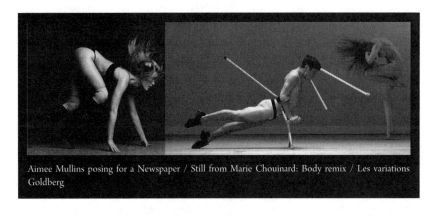

If we compare these two images of prosthetically altered bodies, we get a persuasive impression of western approaches to the body as at the same time natural and cultural: on the left Aimee Mullins is ready predator-like waiting for her dash, while on the right the dancer struggles to move with and against his crutches. Left - a ready to perform 'disabled' person turned into a technically enhanced athlete; right – a,

Aimee Mullins posing for a Newspaper / Still from Marie Chouinard: Body remix / Les variations Goldberg

in terms of his bodily abilities normal or even 'ideal', body, struggling to make sense of the limitations intentionally caused by medical technology.

I want to argue that in Chouinard's performance – but also in images like those of Aimee Mullins – prosthetic fetishes are simultaneously loved and hated 'boundary objects': they don't represent difference and dominance, but they mark the zones within which norms are being negotiated, installed and deconstructed.

I end with another rather allegorical image, which I want to contrast with a still from *Cremaster 3*: it shows Aimee Mullins wearing definitely over length, boot-shaped prostheses watching Hugh Herr, head of the MIT-MediaLab Biomechatronics research group climb a wall with his self-made high-tech-prostheses. The scene took place at the *Human 2.0*-conference on robotics and prosthetics 2007. The group's research is heavily funded by the US Department of Veterans Affairs and

Hugh Herr at the Human 2.0-conference / Still from Cremaster 3

while working at MIT Herr is greatly inspired by ideas of posthumanism. The symposium was therefore dedicated to 'a new era in human adaptability - an era where technology will merge with our bodies and our minds to forever change our concept of human capability' (http://h20.media.mit.edu). The conference explicitly wanted to blur the distinction between 'abled bodied' and 'disabled' – just remember Gilbreth's universal cripple – to promote the benefits of technological enhancements. The *Human 2.0*-conference shows what the debate on prosthetic enhancement currently is about: the fine line between people's right to live their difference, to use whatever artificial technological features they prefer to achieve a good life, good sex or whatever and the will to bring our bodies to an imaginary state of perfection; be it their adaptation to ideals of beauty, be it for the purpose of fitting the body to the machines of production or de-production, to the machines of the military or the machine of economy with its promise of individual and collective advancement. Technofetishism is not so interesting because of its futuristic promise of technologically enhanced human kind but rather because of its ability to reflect on the bio-political and disciplining effects of present naturalized sex and work. As the naive and uncritical fetishisation (Fetischisierung) of a technified future tends to neglect past and present violence – be it physical or symbolic – I would therefore like to propose not the withdrawal from the field of fetishism but to work and think through fetishism's restraints and promises to regain its inventive, critical and liberating side. And just as important: to regain fetishism's recognition for the multiplicity of pleasures our bodies are capable of.

References:

Matthew Barney: The Cremaster Cycle. New York, Guggenheim Museum Publications 2003.

Elspeth Brown: 'The Posthetics of Management. Motion Study, Photography, and the Industrialized Body in World War I America.' In: Katherine Ott, David Serlin and Stephen Mihm (Eds).: Artificial Parts, Practical Lives. Modern Histories of Prosthetics. New York, New York University Press 2002, pp. 249-281.

Manfred E. Clynes and Nathan S. Kline: 'Cyborgs and Space'. In Astronautics, September 1960, pp. 26-27 and 74-75; reprinted in Christ Hables Gray et al. (Eds.): The Cyborg Handbook, New York: Routledge 1995, pp. 29-34.

Michel Foucault: The History of Sexuality. 3 Volumes. New York: Vintage Books 1988.

Donna Haraway, 'A Cyborg Manifesto: Science, Technology, and Socialist-Feminism in the Late Twentieth Century.' In: Simians, Cyborgs and Women: The Reinvention of Nature. New York Routledge 1991, pp.149-181.

Lisa Herschbach: 'Prosthetic Reconstruction: Making the Industry, Re-Making the Body, Modelling the Nation.' In: History Workshop Journal 44/1997, pp. 22-57.

Siegfried Kracauer: Das Ornament der Masse. Essays. Frankfurt a. M.: Suhrkamp 1977. (orig. 1927)

Heather R. Perry: 'Re-Arming the Disabled Veteran. Artificially Rebuilding State and Society in World War One Germany.' In: Katherine Ott, David Serlin and Stephen Mihm (Eds).: Artificial Parts, Practical Lives. Modern Histories of Prosthetics. New York, New York University Press 2002, pp 75-101.

Matthew Price: 'Lives and Limbs. Rehabilitation of Wounded Soldiers in the Aftermath of the Great War.' In: Stanford Humanities Review 5/1996 (=SEHR Supplement: Cultural and Technological Incubations of Fascism).

Anson Rabinbach: The Human Motor. Energy, Fatigue, and the Origins of Modernity. New York, Basic Books 1990.

Richard Kadrey / Johannes Grenzfurthner

SCIENCE FUCKTION
A PUBLIC CONVERSATION

Johannes Grenzfurthner: So, I'm very happy being able to present Richard Kadrey. The last time we met in person was 10 years ago, 1998, and we had a burrito somewhere.

Richard Kadrey: Roosevelt's Tamale Parlor. I remember that.

Johannes: Yes, yes, yes. monochrom was doing lots of stuff back then, and of course the fanzine. I had a copy for you with me, with a short list of forgotten media.

Richard: It was The Dead Media Project, was something I did with Bruce Sterling. It's actually ongoing, but in miniscule way small way. The idea was to try and list every kind of human communication device that is not used anymore. Some of them are very ancient - something like the Quipu, from ancient Peru, to the Singing Telegraph, which is a freakish use of old telegraph gear. You pumped electricity through the device and by adjusting the voltage you could get the thing to hum. And of course, my favorite, The Telharmonium, the largest keyboard instrument ever made. Weighing forty tons and the size of a railroad car, it was also the first synthesizer.

Johannes: I read a nice story, that back in the 1980s, someone told me, 'You know, the computing power of the whole Apollo program, bringing a man to the moon, that's pretty much the Commodore 64.' I was like, 'OK, wow.' Commodore 64, but nowadays what shall I say it's like; it's a tiny little part of an iPhone? I don't know, it's like a Commodore 64, it's a wristwatch.

Richard: They had three 64K computers in the Lunar Module and all three failed. Armstrong landed it by hand. So much for trusting your tech to keep you safe.

Johannes: The Americans invested quite a lot of money into creating a pen that can write in zero gravity. The Russians, of course, always used pencils.
[laughter]

Johannes: But then again, there's a problem with the Russian side, because you have to sharpen the stupid pencil. And if that fucker opens, oh my, that would be really bad. Maybe, that's a good start for talking about Science Fiction and sexuality. [laughs]

Richard: Yes. Clever tech constructs that seem like a great idea, but never quite get you that happy ending.

Johannes: Richard, you are a BDSM photographer? You are a science fiction author, you are a fantasy author and that's why I think it's a really good idea to have you on the stage here. Because at *Arse Elektronika*, the topic is: do androids sleep with electric sheep? It's about sex and science fiction and social fiction, of course.
Could you tell me a little bit about your, in German we would say 'Sozialisation'? How did you start writing science fiction and how did you start getting interested in BDSM and do they work together?

Richard: There's a very logical connection between modern science fiction and the fetish world. In both you're talking about the fetishistic object. In SF, there's a kind of obsession with the form and functionality of the future, of the objects we use in the future and how they relate to us and often shape our perceptions and behaviors.

Fetishism often works the same way, the obsession with the object and your relation to it. A simple example is foot fetish, which encompasses shoes and high heels. The shoe is the object of desire. Not the person wearing it. It's the obsession of surfaces. Fetishizing the tech that is taking us into the future, the iPhone, netbook, RealDoll and Second Life.

Johannes: It's very interesting that a couple of really good, science fiction collections were published by Playboy, in the 1970s.

Richard: Playboy was the first American publisher of Ballard's Crash. They did the first paperback. The cover was a sports car with tits for headlights. I lost my copy years ago. It's an amazing cover - beautiful in this complete, trashy, exploitation, pulp way.

Johannes: So we have the guy, called JG Ballard, and he writes about techno-fetishism. He writes about sex and death. Here at *Arse Elektronika* we have a machine designed by Stephane Perrine. In one mode, this dildonic device would only vibrate if there's an earthquake somewhere on the globe. And in the other mode it measures Iraqi civilian body counts. I like this connection between sex and death and obviously, sex and death is a major theme going on in Crash.

Richard: Crash is the one book everyone points to, it's the most obvious one, that real obsession between machine-body and the destruction and rearrangement of both. But there was plenty of stuff before that. Doctor Adder, by KW Jeter, is a book about this radical surgeon living in a future LA. He finds the sexual obsessions of willing subjects and then alters their bodies in these radical ways: amputation, augmentation of genitalia. He turns people into organic fetishistic fuck machines.
It was only published in the 80s, but it was written in 1972. It was only published in the 80s because no one would touch it in the 70s.

Johannes: The early 70s seem to be an interesting time. One of my favorite short stories about sex and science fiction by Mister Tiptree was published in 1971. Actually, Mister Tiptree was no guy; it was a pseudonym for a female science fiction author.

Richard: Alice Sheldon.

Johannes: Yes. She wrote an excellent short story about space stations somewhere in the middle of nowhere. She's writing that humans are pretty much completely obsessed with aliens because they are obsessed by them in a sexual way. The aliens are not forcing them to work for them or do like shitty jobs for them. But the humans want to be near the aliens, they're so attracted by these alien creatures and they want to fuck them.
She even talks about the colonialization of Polynesia and how the intruders literally fuck up the genetic order of whole islands. But intercourse with aliens never can fuck up their genetic order. We cannot conquer them on a genetic way. It's almost like moths burning on a lamp. A basic drive the moth can't withstand – kills it.

Richard: There's a long history in science fiction of that obsession with the Other. It's a logical extension of meeting the aliens and it goes back a long way. It might be something hardwired into us. I've read myths from all around the world, and in every single myth system someone is having sex with a god. Usually with bad results.

There's a fairly romantic machine sex story by Lester del Ray from the late 30s called Helen O'Loy, in which a man and a friend of his construct a female robot. He just falls in love with her in a completely romantic way. Basically the man and the robot were this husband and wife for their entire lives.

So that obsession's been there for a long, long time. It's not explicit in something like The Time Machine by H.G. Wells, but there are certain fetishistic implications in the book. You have the underground dwellers, the Morlocks, and you have the Eloi on the surface. The Morlocks are constantly kidnapping the Eloi and taking them down below into their dungeon-like caverns.

Johannes: And consuming them...

Richard: And consuming them, which itself is a powerful sexual metaphor. Combine that with the image of these brutes storming in and carrying away these weaker humans, women and small, feminized men. This is like every kidnapping and fake rape scenario ever played out in a fetish club. And you can take it further. There was Armin Meiwes, the guy in Germany who put an ad on a website asking for another man who wanted to be killed and eaten. A volunteer showed up and they both went through with it. That's a JG Ballard story come to life. Much crazier than Crash. Sexualized cannibalism meets the global communication web. Do you think the people who built ARPANET ever thought of that scenario?

Johannes: Nowadays where random porn is pretty much accessible on the Internet and it's pretty easy to produce. I mean, tentacle porn is harder to shoot, but…

Richard: I have all these friends who are now obsessed with tentacle porn. Porn star Mandy Morbid actually did a tentacle porn shoot with a monster they built for the film.

Johannes: I thought that tentacle porn only exists because of the Japanese censorship thing going on that you can't show a dick and therefore someone started to create tentacles. But I think it's only an urban myth, and the sex and tentacle thing has been going on in Japanese culture for a couple of hundred years now.

Richard: You can see it in old paintings; it's long-term obsession. Japan is an island nation and they have this very strong relationship to the sea. A lot of cultures that have that strong relationship to water have similar mythology. There are Amazonian myths about freshwater dolphins, and in fact supposedly there are these coming of age rituals where young fishermen fuck the freshwater dolphins down there.

It also works for the women because sometimes the dolphins come out of the water, and fuck the women along the Amazon. Then they have these alleged merbabies, which is a really nice way to cover up that you're pregnant from an affair. You're just sort of like 'it was the dolphin, honey. Not my fault.'

Johannes: There's a nice Oceanic myth: that the world was created because one of the gods fucked an anthill, I mean I really like that. I probably wouldn't try it, but he is a god, anyway.

Richard: Depends on the ants, I suppose. Some ants are sexier than others.

Johannes: Many, many science fiction films and novels are extremely violent. So they're in a certain way segregating what's going on. So they're finding metaphors, and amplifying the real world. So am I right to state that science fiction is never about the future, it's always about the present.

Richard: Most of it because anyone who tries to predict the future is clinically insane.

Johannes: I don't believe in the concept of future. Where is it? We are stuck in an eternal state of the present. What really interesting examples of sex in science fiction are there in the last couple of decades?

Richard: What's really interesting over that period is we had this boom in the 70s, and then it slowly died down again - especially in science fiction. I can't say if that's marketing forces, or if that's just a general repression that began in the 80s. There was just a lot of fear during the Reagan years that was blowback from what a lot of middle class Americans considered the hyper sexualized 70s.
But to go back to the 50s, the 50s are really interesting because in that post war world was consumed mostly by fear. You have the end of a war. You have the rise of communism. The Soviets are becoming stronger and they have the bomb. There's this overriding fear that runs through the whole period. On top of that, there's all these men have returned from the war and expect to go back to normal life and jobs. But during the war, women took over a lot of these very manufacturing important jobs. While the men were gone, women moved up the food chain and there was this brief, proto-feminist period with women doing these heavy industrial jobs, building machines, building jets, and building bombs for the war effort.
Then, all the men came home and suddenly it's 'thanks for building all those tanks, honey. What's for dinner?' And most women went along with it because there was a desperate desire for a lot of Americans to return to what they thought of as normal American life.

Johannes: And especially because they were thinking 'what's the normal American life?' They were returning from the war, and pretty much fictionalizing the whole war about 'I want a normal life.'

Richard: They pretty much idealized what we look at now as the 'Leave it to Beaver' world, which just became some imaginary goal while you were living in filth, and shit, and being shot at all the time. So you have this return to a fantasy version of normalcy, with this rigid and codified gender role system. I think the easiest way to talk about it is one of my favorite films is Them, the giant ant movie from the 50s. It's actually a pretty good movie.

Johannes: Them! Oh yeah, great.

Richard: Talking about sexual roles and things like that, it's a perfect one that covers a lot of what happened in the 50s. Essentially giant ants are created by a nuclear test, that's all you need to know.

Johannes: It's a film noir about giant ants, and that's quite a statement.

Richard: OK, let me just say this, and I'll argue why it's not noir.

The social setup in Them is great, and it's something that carries through all of the 50s. There's an older man, an Einstein stand-in, a sort of genius. But he's a completely desexualized figure. In an old myth he'd be the wizard that knows everything, but has no relationship to the body.

In Them, you have the three main 50s archetypes. There's the Einstein-like scientist, his daughter, and the outside guy who comes in - the American man of action. He's thrown in with a desexualized genius and a dutiful, sexually naive daughter - a vestal virgin, waiting to be swept away by the rugged American uber-business man.

Johannes: The first thing you see of the daughter is her ass. She is climbing down a ladder of a plane, close-up. And then you learn she is the scientist.

Richard: She is a scientist but she is not allowed to be an adult woman. That's not her role.

Johannes: But she is allowed to climb into the giant ant hill, she is even allowed to burn it.

Richard: Well no, she can do the kind of science thing, but she cannot react at all to the sexuality of being an adult woman. She is just taking care of dad and doing science.

She doesn't have any relationship to her body or sexuality until the other guy comes in and that is the man of action. That is the dude coming back from the war. That is the cowboy. That is the guy who is going to rescue her from the ivory tower and fuck her to life.

Johannes: It is pretty much like the father figure. It is like this baby boomer generation father figure in a certain way - the guy coming back from the war who can handle it.

Richard: The virile guy coming back. The protector - a daddy figure with a dick who will magically transform her from a woman with possibilities into a housewife. The American dream. That is the end of that story that they don't show you in the movie. 'I know you are a scientist and you saved us from the giant ants, but make dinner.'
[laughter]

Johannes: Let's talk about robots.
[laughter]

Johannes: But then again, robots are quite asexual in the 1950s. And if you go on to the 1960s, there is crazy stuff going, like Barbarella. I am not even sure if Barbarella was a financial success.

Richard: I think it was, yes.

Johannes: Sex and ticket sales?

Richard: Jane Fonda, I mean, Jane Fonda at the height of who she was, too.

Johannes: It is a European film so...

Richard: That is right. Roger Vadim directed it.

Johannes: Could you think about a specific, a really important movie or book of the 1960s, that is somehow addressing sexual roles? Maybe not in this pseudo-provocative way as Barbarella does.

Richard: Before it was a movie it was a French comic. I think it was in - what is the French name for 'heavy metal'?

Johannes: Many people are referring to Barbarella as the epitome of sex and science fiction in the 1960s. But the early 60s were pretty conservative.

Richard: It is in the 60s that you first start seeing this sort of new idea of social roles. Where you start seeing things like redefined gender roles. Not such rigid divisions between gay and straight, what kind of marriages are, but you start seeing the idea of what our future families. You are starting to see things like group marriages. People don't really think about it now, but Samuel Delany started publishing in the early 60s and moved into that area very quickly.
I think even something as dicey as something like Stranger in a Strange Land is coming out of that culture and from that period. I don't know if it was actually published in the 60s but it comes out of that kind of ethos, even though it is a very questionable book in some ways. Heinlein was an interesting figure in that, in which he kind of vacillated back and forth...

Johannes: He was a polyamorous guy, actually.

Richard: Intellectually he really was kind of all over the place. Some people kind of excused some of it because apparently he had some kind of brain blockage for a lot of his life, and he was sort of deranged.
He was an older guy who came out of a certain culture and so when he was trying to write kind of radicalized stuff, he was still at the same time very homophobic.

Johannes: I am stepping away now from the time-line but I like what Verhoeven did with Starship Troopers, when he filmed it in the 90s. He takes it and twists it around and creates this antifascist metaphor and spits it out into pop culture. Most of the people who went to see the film didn't even get the message. People were just thinking - oh wow. Cool young folks fighting insects. But what they really are is mediocre fascists.

Richard: It's a fascist romp. It is my favorite thought about the thing, which in the end becomes this anti-fascist thing. But it is this beautiful banal representation of all that kind of ethos. That again, isn't the future. The Nazi imagery isn't the past; it is definitely that kind of rigidity you are seeing in, again, George Bush's America.

Johannes: In the 1970s, there is lots of disturbing things going on. I mean, Star Wars itself is a highly dystopian Jesus flick.

Richard: I don't want to talk about Star Wars. Fuck Star Wars.
[laughter]

Richard: Actually, Star Trek has some very weird gender stuff, if you want talk about that.

Johannes: The only thing I am always thinking about Star Trek is, do they teleport the shit out of people? They don't have toilets. There is not a single toilet in the whole fucking Star Trek universe.

Richard: I actually wrote an article about that once - all the back room stuff that they don't talk about on Star Trek. Everyone's drinking everyone's piss. If they're out there for a long time, you're recycling all that water. And the great part is: you're drinking alien piss! So, who the fuck knows what that is.

Johannes: OK, the 1970s dystopian world...

Richard: It's not necessarily dystopian. The 70s are when science fiction officially went batshit.

Johannes: For example Flucht ins 23. Jahrhundert…
[laughter]

Johannes: I don't know the English title. Michael York was the main character.

Richard: Logan's Run?

Johannes: Of course, such a nice and short title - Logan's Run!

Richard: Please translate the German title - what does that mean?

Johannes: Escape into the 23rd Century.

Richard: What an awful title! Terrible title.

Johannes: Brrr. German dubbing. They did a lot of bad things.

Richard: Actually, may I ask you a question? Because I know in some other countries, what they're doing with science fiction - not any more necessarily - is because it's just genre material, they just gave it to some guy and said, 'translate it in 24 hours' - it's just crap.

Johannes: Book translations are usually quite good. There's a big publisher in Germany - Heyne. They were bought by Random House. They're pretty OK.
Sometimes the titles get longer, and sometimes the titles get shorter. There's Down and Out in the Magic Kingdom by Cory Doctorow. The German title is Backup.

Richard: It's a very retro title for the 21st Century.

Johannes: Heyne called it Backup because they want potential buyers to know it's something about net stuff.

Richard: That's a good title - 'Something about Net Stuff' - let's call it that.

Johannes: [laughs] Back to Logan's Run. There's definitely a sexual component in this hedonistic world, where you actually can only live until you're 25, and then they send you to the Carousel to burn you.

Richard: It's all sex. That's all that movie is really about, because it is that worship of youth and beauty. Whatever else is going on, it is purely a sexual theme. And death - which is the inevitable end of the society's decadence. Not only does society ritually kill you, but there is a whole separate class of people who kill all the time instead of fucking all the time. It's Freud with nymphet fashion models.

Johannes: It's like at the Google campus. I mean, when you go to the Google campus, the whole architecture reminds you of Logan's Run. They even have free Naked Juice there. The whole architecture of the Google campus is highly reminding me of...

Richard: A little bit like Starship Troopers.

Johannes: All the people sitting around there, working and programming, playing Billiard and drinking free Naked Juice and having a whirlpool there, feeling fine. When they're 25, they probably get burnt.
[laughter]
Or they have to work for Yahoo. I don't know.

Richard: Well, you do burn out in those places. I did my dotcom time. It was kind of fast and there was this fake hedonistic part of it, and these fake comforts. Yes, there's a whirlpool. Yes, there's a couch.
By the way, you're working 80/100 hours a week. It's this fake kind of payment for you. At least in Logan's Run they were getting laid.
[laughter]

Johannes: I don't know about the 'getting laid' count at Google. I don't know about that. But I know that they have the Google shuttle bus, taking Google's employees from their homes to Google. They have, of course, wireless Internet access in the shuttle bus. So the Google employees start working when they step out of their home into the shuttle bus. Actually, they don't force them to work. It's just like, 'hey, we have access here, so you can use it. Do whatever you like.' And of course...

Richard: That's sort of the anti-sex part of the modern world, that doesn't get talked about that much. The future Logan's Run is all about hedonism, but the result of our technology has often been the opposite. Work starts to encroach on our regular lives and to take us out of all the leisure time that we were promised at the beginning of this.
There's a really good example at the beginning of the 20th Century. There were all these household products made for women - the vacuum and especially the washing machine. Before that, women had this drudgery of once a week laboriously cleaning a pile of clothes the size of a bear.
It was terrible. So, what did we give women? We give women the washing machine, to help them out - to give them leisure time. Only, what really happens is that now the family expects to have clean clothes 24/7. So this lovely washing machine that's supposed to free women from the drudgery of washing, consumes their lives.

Johannes: Of course, there's the shift from a disciplinary society toward a society of control. In a disciplinary society, you pretty much know who your enemy is. There were workers and there were bosses. There was a clear antagonism between the worker and the boss. There even was a whole workers culture constructed around that. And they pretty much knew the boundaries and how to attack and subvert them. Nowadays, in the society of control, as Deleuze pretty much describes it, the boundary is blurry, you are a friend of the boss, and you feel free, part of the big corporate family. That's what's happening at Google.

The hierarchies are getting flatter and flatter. But there's still the same division between the workers and the bosses who own the means of production. It's still the same bullshit going on. You just believe that you're in control of your life. People feel free, and suddenly they are their own worst enemies.

There are good science fiction novels out there addressing this shift of going from a strict dystopian, authoritarian regime to new regimes of soft control.

Richard: Look at Douglas Coupland's Microserfs to see that world. It's supposed to be modern tech world social fiction, but to me it's always read like H.P. Lovecraft - digital culture horror.

So, now we're up to the 70s. The 70s are when things just went batshit. That's when you have Ballard, New Worlds in England, and the New Wave in the US. Basically, the shift in the 60s up through the 70s, was rejecting the idea of the old science fiction rockets to space. The 60s and 70s became about inner space. That was one of the major changes.

Johannes: A New Wave, in a certain way.

Richard: That's when Ballard published The Atrocity Exhibition, micro-length, hypersexualized stories that could have never been published before. Let's plug Vale and Re/Search. He has a very good illustrated edition of The Atrocity Exhibition that you should all buy and take home and love.

During that same period, you have fucking crazy people, like Philip José Farmer.

Johannes: Oh my.

Richard: He's the guy from that period. I don't know where this guy came from, because he started out as a fairly straight science fiction writer. It was almost like someone said, 'Phil, go!'

And he started writing the craziest novels. Flesh was my favorite one when I was young, just in terms of the basic premise. An astronaut goes to the other planets, has antlers surgically attached to his head, and then just starts servicing an entire world of women.

Then it gets even odder, because he starts playing with old pulp conventions, dragging them up into this very sexualized fiction. He rewrote Doc Savage, a big popular pulp figure, in very violent and sexual ways. And then wrote a book in which Tarzan and Doc Savage meet and basically rape each other.

He's brilliant, if you can find any of those old books. Just anything by Philip José Farmer, from the 70s is worth reading.

Then you also have some serious stuff too. You have Johanna Russ, the first serious intellectual attempt to bring this lesbian, female-oriented kind of science fiction. The Female Man is a really good example. It was basically the feminist lesbian answer to Starship Troopers.

A very, very different book than anything anybody had ever seen before. Again, Tiptree is coming through that period. 'The Screwfly Solution' is also a really important story. Who knows the publication year?

Johannes: One sec, checking it on my iPhone. … 1977.

Richard: Thank you. That's a really amazing story. It's maybe the single most depressing story about men and women in science fiction; terribly depressing, but brilliant. There is a virus that appears that causes men attempt to destroy all women.

Johannes: That's not so far off, I guess, or?

Richard: Well, again, in George Bush's America, anything is possible.

Johannes: I am from Europe, and there's quite some patriarchy going on there too. So, I guess we don't have to diss the US of frickin' A too much about sexism. Or should I?

Richard: Feel free. We're Americans. We're not listening.

Johannes: The free speech thing. I don't believe it. I don't know. Freeish speech, yes.

Richard: Freeish. We're a constitutional-ish country. You don't want to go too far with this 'freedom' thing.

Johannes: We are pretty much heading up the road to the 1980s. There's, of course, stuff going on like The Handmaid's Tale.

Richard: It's a very prescient book. Again, going into really what's come. People who lived through Ronald Reagan, you didn't think things could get more fucked up. You didn't think things could get worse.
I remember the entire Reagan administration, living in LA, waiting for the bomb, waiting for the thought police to come and take away all your books, all your videos, all your records.
The Christian thing really started bubbling up through Reagan. The religious right had been plugging away before that, and Reagan was their first success, on a national level. They'd been doing really well on a state and local level up to then. But Reagan was the first national figure to really carry that religious right message to the whole country.
The Handmaid's Tale is a very logical extension of what was just starting. It just seems bizarrely prescient when you consider where we've gone since then.

Johannes: Iran is not a free country. There's lots of bullshit going on there. I mean, homosexuals are being killed there, just for being homosexual.

Richard: They're being killed here, too. Go to Texas.

Johannes: Oh my.
It would be highly interesting to combine the main theme of The Handmaid's Tale, and overlap it with the current discussions about political Islam.

Richard: The fear now has shifted from that 50s thing of the bomb, the Soviets, very specific things that you could point to. Stalin was scary. Eastern Europe was scary. What freaks people out now is it's scary, but in this very amorphous way because we're not fighting a country. You're not really fighting Iran. You're not really fighting Iraq. You're fighting an ideology, which covers the whole globe. I think that's where the generalized paranoia comes from.

It's a better paranoia for people in power because it's endless. You can conquer a country, but someone will always believe in an ideology, no matter how fucked up it is. Someone will always be willing to strap on a bomb and walk into a marketplace, so the fear is endless. A perfect excuse for a police state.

Johannes: It's like the Forever War, Joe Haldeman's book from the 1970s where he's talking about the Vietnam War. It's a pretty sexual book, actually. I think in the end, they have this completely homosexual semi-utopian society.

Richard: It's at times like this, because of generalized paranoia, we love to give up rights. This illusion that we will be protected, if we just let others take control of our lives. That's when again, in this more subtle way than the Reagan administration, you do have this war on thought, especially sexual thought.
Everybody I know in the porn biz is unbelievably paranoid now. For instance, everyone has lists of places and people you cannot sell stuff. There's a guy I know who runs a gay porn company in the city. He has very rigid rules about how he sells his products. He will not to some states. He sells to jobbers with the agreement the jobbers won't even sell in certain places because they're set ups. There are literally entire states that are set up to get you to mail something to them so that they can then prosecute you.
You have prosecutions of a lot of pornography now; you have the Max Hardcore case, which is really a big one, where they finally sort of found somebody who bothers enough regular people, even enough regular porn people. They found the guy they can go after and prosecute and seem to be winning. It's a great time to just chip away at everything, especially things that scare the religion right. And there's nothing scarier than sex to that part of the world. Growing up partly in Texas I learned when someone talked about the Devil, it was not metaphorical. It was as literal as could be.

Johannes: That's why I like Protestants; I was raised Catholic, so at least we can confess, but the Protestants… they're just completely screwed…

Richard: You put something in the plate at the end of the service. You just pay money. Basically you're buying indulgences; they just don't call it that.
I grew up; when we went to the local churches in Texas, they actually would publish how much money people gave. That's the kind of weird pressure these people would put on you in the South. They would literally have lists in the back of these church bulletins of who was the cheapskate.

Johannes: My favorite Simpsons quote is by Homer Simpson: 'God is my favorite science fiction character.'

Richard: There you go. But that ethos we're talking about… that really started coming into bloom in the 80s, when a lot of stuff, a lot of the sexual part of science fiction in the mainstream, really started tapering off.

Johannes: And that's pretty much like the nutrient fluid your science fiction grew out of, I would say.

Richard: I was completely a New Wave kid. I hated most 80s science fiction. I hated that stuff. It felt like sort of fluffy, a soft return in a lot of ways to the older ideas of science fiction. Again, probably in reaction to the New Wave.

Johannes: When did you start actually writing science fiction?

Richard: My first book came out in '88. My first stories are about '86, or something like that

Johannes: Could we like sum up the round because we have been talking for over an hour now? Could you - how would you explain your relationship to science fiction and sex? Your personal relationship?

Richard: Ironically, I am not writing science fiction very much very much these days. I'm writing fantasy for the simple reason that I can't make any money writing science fiction, so I'm writing other stuff.

Science fiction has always had a troubled relationship with sex. You have a period of the late 60s and the 70s when it opened up, but that's gone much down again, except in small press SF. However, in fantasy, there's this explosion of sexual stuff. They are entire lines now of fantasy novels that are sort of fantasy romances in which, basically, nice ladies and nice guys fuck vampires. Some supernatural romance novels are pretty much porn with werewolves, vampires, whatever. That, in the genre world, is where sexuality has gone.

The only place you find science fiction now, which is kind of depressing, is specifically in places where they're saying, let's write about sex and science fiction. You find little, small press anthologies, you have gay anthologies, lesbian anthologies, lesbian stories about sex and science fiction, gay guys' stuff about sex and science fiction. You don't see it very much in the mainstream.

There are a handful of writers; Samuel Delany, is probably the classic guy who, through this whole period, has just ignored all the rules all the way through it, and wrote some amazing stuff through every generation.

Dhalgren is one of those books that if you were a kid wanting to write science fiction, it just blew your mind. You hadn't seen anything like this before. It wasn't about sex, it was simply that the sexuality of these characters was just part of who they were and it was so just thrown away that someone was gay, someone was straight, someone was bi. None of it was the point of the story. It was just how they traveled through this world.

He's done that same thing all the way through his career, exploring different roles, but not any kind of showoff-y kind of way. It's just, 'Oh, here's how things are.' 'Aye, and Gomorrah' is another one of those stories the 70s that's still amazing. It's about humans fetishizing astronauts who have to be neutered before they can go into space. Like the alien in the Tiptree story you mentioned, the astronauts become fetish objects when they come back to Earth.

Richard: Angela Carter's great book, the Red Riding Hood anthology, The Bloody Chamber, rewriting of classic fairy tales, almost bringing a modern Brothers Grimm back to it and adding a female element, along with dangerous sexual themes.

Audience Member: You kind of implied sexuality is more in fantasy right now that in science fiction. What - do you feel sexuality is more accepted in fantasy right now than in science fiction? What do you feel is different about fantasy that allows that's not allowed in science fiction?

Richard: I think in fantasy, it's very easy to have those fetish objects. Dracula started off as a real fetish book. I mean, that's the one that really hyper-sexualized the idea of the vampire and it has never stopped. And now they've made it into this pop product. They've added some other, you know, werewolves and ghosts and other things to it.

But, it's also very simple to say vampire - you know what a vampire is. You're not explaining anything like in science fiction - you just, 'want to fuck a vampire'. It's there.

Audience Member: Hi, I was so glad you mentioned Delany because my head was ready to explode. I was going, 'Where's Delany here?'

Richard: No, he's one of the...

Audience Member: And, I would also say, Octavia Butler. I mean...

Richard: Yeah, Butler too. Yes, absolutely. She was one I kept thinking of, but I never quite got around there.

Audience Member: Yes, yeah, I mean she does human alien sex and reproduction and miscegenation just about better than anybody.

Richard: Similar to the Tiptree thing where it's very complicated and often it's not the fun kind of fetishizing of the alien. It's very complicated and very often tragic.

Audience Member: I want to say something about pornography, too, if that's not out of order.
The Max Hardcore prosecution is very scary. He is someone who does stuff very much on the edge. He's like a bad boy of porn; the way that Quentin Tarantino is a bad boy of film or Jeff Koons is a bad boy of art.
I find even scarier, is John Stagliano's prosecution. He was just indicted on seven obscenity charges; this is Buttman. I teach a class on pornographic film, and I invited John Stagliano to come to my class and show the indicted videos. Because of course, that would give them scientific value, showing them in my classroom at the University of California, Santa Barbara.
But what was really shocking about the videos is that... they aren't his films. They are films made by people in his company. They are not soft; it's all women exchanging bodily fluids and then a little bit of light bondage.
So, they're going after... This is what the Right has wanted to do. Don't just go after the really bad stuff. It's all bad, so we want you to go after the more mainstream stuff, too.

Richard: Again, you have Max Hardcore, a scary guy with a scary name. Then you have Buttman. You have somebody who is... Again, that sounds like anal sex, which sounds like the worst thing in the world to a certain part of the country. I heard a conservative family values spokesman giving a talk about his relationship with his daughter and how he wants to protect her now that she's in public school. It was pretty much 20 minutes on his daughter, lesbian seduction and anal sex.
[laughter]

Audience Member: There was a lot earlier of talk about fan fiction.

Richard: I think we have to talk about slash fiction and stuff, because I think that's the real modern area for sexuality in SF

Audience Member: Yeah, well I was going to ask if you feel that there is a relationship between the growth of that movement and the decline of sexuality in more traditional science fiction.

Richard: Sadly, some of the conservative writing in SF comes out of basic market forces. Sexual romance fiction is kind of covert. No one acknowledges what it is and the audience is older women. But for book chains, it's teenagers that read SF. That makes sexuality more dangerous and can it can make it harder to get your book into a Barnes and Noble or Borders. If you're not in there, your book is dead. The chains control an enormous amount of the market and have a strong editorial voice when it comes to what they will and won't sell.

I think also think that slash fiction is an amazing thing. In a lot of ways, slash fiction really is much more science fictional than most science fiction right now. It's like mash-up in music. Basically, you are completely appropriating media objects and remixing them in this very personal way. I think it's charmingly radical. There are a few academics, now, that take it very seriously.

Look online. Whatever it is you like, you will find slash fiction for it. A friend of mine, who's obsessed with Buffy the Vampire Slayer, was showing me reams of this stuff, of every combination of those characters in relationships that are everything from titillating to really hardcore S and M behavior.

Johannes: And we are talking about Star Wars.

Richard: We made our way back to the dreaded Star Wars.

Johannes: Yes. [laughs] He actually didn't want to talk about it.

Richard: If you're going to deconstruct Star Wars and completely ruin it for regular Star Wars people, I'm on your side. Those Star Wars people. I was dating a Star Wars fan when the last part of the trilogy came out. And I was literally not allowed to speak for a day, because she knew what I would say.
[laughter]

Audience Member: On slash writing, the quality of slash writing, Joanna Russ who has done some slash... Yeah, she has done some slash stories, and she told Marge Percy, who told me, 'forget Bread Loaf; forget the Iowa Writers' workshop; Slash Fans is the best writing workshop around.'
[laughter]

Richard: There you go.

Johannes: There is a certain competition in fan fiction.

Richard: Well we need to take it further. They want to impress each other; they want to go more and more extreme and crazier and good for them.
I think it's out of slash fiction that we're going to have the next Delany, the next Philip Jose Farmer, just this crazy, crazy writer who has his or her own vision and just does not care about what the rest of the world or what the rest of commercial fiction is about. That could be a good place to end.

Johannes: OK. Yeah. Maybe a last question and then we should probably stop. OK, yes, of course.

Audience Member: Earlier, when you were talking about Them… You were kind of gesturing toward the Lacanian dynamics, which I feel is pretty typical across a lot of genres.

The fiction is not really specific. But, what I was thinking about was the maternal thing that goes on behind the daughter and her daughter and father, and how that plays into her role as a sex object and things like that. I was wondering if you could talk a little about how those themes of maternity wind up playing into science fiction, and if they function in the same way as when there is a world established that doesn't even necessarily have biological reproduction in the way that we think about it.

Richard: That's the woman's role in so much, really, not even just science fiction - in pop fiction, genre fiction and action fiction. The woman is there to take care of the man of action, to be the person you go to for stability, but with nothing else to do in her life. Again, even if she was the lady scientist, there is nothing else there.

That's just pulp fiction. That goes back as far as you can see any popular fiction. The variation on that is the kind of simple, quiet maternalistic - or the Electra complex - character, who then becomes kind of sexualized. She is the object of desire by the alien. You can see it on the old pulp covers.

But that's just a pulp replacement for the woman being wanted by the foreigner. If you go back and look at the 20s, it was Chinese. It was the Fu Man Chu who wants our women. Before that, it was all those scary black men, freed slaves. Starting in about the 30s, it was little green men.

Yeah, it's all the woman is the object of desire, but not having her own sexuality. She is very closed, always representing the home, always nurturing, no matter what other role she may have had. No matter whatever other role she may have: She may be a scientist now, but she's still the homemaker. It's still there, in a lot of ways, especially I think in film, where everything gets simplified and codified.

Johannes: Thank you very much for being on the show, Richard.

Richard: My pleasure.
[applause]

Richard Kadrey

TREMBLING BLUE STARS

I was sitting at the counter, drinking espresso and smoking Gauloises at the Hellas Basin Cafe on Rozhdestvenka Street in Moscow.

The day before, we'd been riding the veer, ferrying supplies to an ASEAN research facility deep in the Oort Cloud. It was pleasant to be back on Earth. During each veer run, when time-space turned psychotic and the heavy rad poured in, we would go null and let our guests do the driving. These petit morts moments were necessary for deep space travel. Dying wasn't such a bad thing if you knew that cigarettes and strong coffee would be waiting for you when it was over.

A woman walked up behind me and said, 'Those black lines across your knuckles and the backs of your hands. I know what those tattoos mean.'

'Do you?'

'You're a cosmonaut. A deep space cowboy who rides a twenty-kilometer bucking bronco between the stars.'

'Clever girl. Your parents must be very proud.'

'Hello, Arkadi.'

I'd seen her when she walked in. My guest's oddly augmented vision revealed her in a distorted panorama of the cafe the moment she entered. I was hoping she wouldn't see me.

'Hello, Valentina,' I said and turned my head politely in her direction.

'I thought you were dead.'

'I am.'

'I know. That was a joke,' she said.

'A good one, too.'

'You look very handsome for a corpse.'

'Thank you.'

Valentina had changed since I'd last seen her. Her hair was long, well past her shoulders where before it had been buzzed close to her scalp. She wore little makeup and was noticeably thinner, but no less beautiful. Her nails were short. It looked like she'd been gnawing the ends.

'I didn't really believe that you'd go through with it. You let them murder you so you could be a handbag for a parasite.'

'Don't talk to me about that tabloid nonsense. It's the newsfeed's way to sell ads. Aliens and ghosts roaming the stars slots in nicely between celebrity gossip and government conspiracies.'

'But you are dead, Arkadi. They took out all your organs. You're as hollow as a chocolate Easter bunny. You don't even breathe. That thing inside you feeds you oxygen.'

'Our brains still work. As long as our brains work, we're still ourselves.'

'Are you sure? I read that the doctors do things to your brains.'

'It's just a volume adjustment. Our guests can't stand all the noise, the stimulation in ordinary human brains. But we're still who we are, just a bit steadier.'

'You let them lobotomize you.'

I turned fully toward Valentina. It was purely for her benefit, to give her the sensation of personal contact. She was leaning one elbow on the counter looking at me that way a poor wife might consider the last, sad chicken in a butcher shop.

'This is a very interesting conversation,' I said. 'We haven't seen each other in over two years and all you want to talk about are the whereabouts of my liver and pancreas.'
'I want to know that it's you I'm talking to and not some meat puppet run by a space monster.'

I drank the rest of my espresso and held up my cup to the waitress for more.

'How did you know I was here?'
'I didn't. I eat lunch here two or three times a week.'
'Why would you eat at a cosmonaut hangout?'
'Didn't you hear? My husband left me to run off to play spaceman with his friends.'

She reached across the counter and took one of the Gauloises, picked up my hand and used my cigarette to light hers. The subtle differences in our ambient skin temperature made her hand feel very warm. My guest curled itself inside me, retreating from the sensation.

'How do you smoke, if you don't have lungs?' she asked.
'I contract and relax the muscles of my diaphragm. It takes a little practice. The good news is that I won't ever get lung cancer.'
'I fucking hate you.'
'Good' I said. 'Then you can let me go and forget you ever knew me.'

A couple at a table by the window got up and left. Valentina took me by the hand and led me to where they'd been sitting. Through the window, we could see the outline of the Kremlin and the old Savoy Hotel, a bright pre-Soviet bauble, built like a ridiculous toy fort, a fantasy castle for a giant child. Its old-world lines were marred by a patched mylar dome on the roof hiding an array of microwave antennas, satellite relays, water and air scrubbers. In the plaza below was a traveling carnival with glowing, swooping rides and virtual wild animals prowling the grounds.

Nothing stays the same. The old is jettisoned. The new incorporated. Everything, if it exists long enough, is a chimera.

'How could I forget you when you just disappeared? You could have stayed long enough for me to throw you out. That would have been the polite thing to do.'
'I'm sorry. I wasn't thinking very clearly at the time. I was scared. I was scared of you; and I was scared of wanting so badly to go to the stars. I'm better now, though.'
'And all you had to do was let them gut you like a fish and fill you up with an alien parasite. Were you really that anxious to get away from me?'
'The alterations are necessary. Humans can't function in deep space. Our guests are the only thing that makes it possible. It's a symbiotic relationship.'
'I think you love your alien more than you ever loved me.'

I gestured to the waitress to bring us our coffee by the window. I drew on my cigarette. A pleasantly acrid stream of smoke filled my mouth. Cigarettes are the perfect prop when you have nothing to say.

'I didn't stop loving you. I just loved the stars more.'

'But you don't love me now.'

'I can't love you. They cut those kinds of things out of me.'

'Right. Cosmonauts are heroes and above all those sticky human connections. What do you suppose they do with your excised brain matter, Arkadi? All those feelings locked away in the dead neurons? Do they recycle them into pet food? Ritually burn them and throw the ashes in the Ganges? Maybe they give them to the aliens as trophies. 'Look at the shit we can talk people into.''

The waitress bought our espressos and Valentina drank hers looking out the window. Coffee is another good prop for when you run out of words.

A light snow dusted the street, the first hint of what was supposed to be an especially cold winter. Valentina was lost somewhere in her head. I took the moment to do a light meditation. I went through the number patterns we recited before heading out into the veer. A Fibonacci sequence. Cubes. Twin primes. The steady sequences calmed my guest, who'd grown restless from the moment I'd laid eyes on Valentina.

She finished her cigarette and dropped the butt into the dregs of her coffee.

'You should have taken me with you,' she said.

'You know I couldn't.'

'Deep space is a boy's club. Girls can't play.'

'You can't blame me for that. There are basic biological incompatibilities between female neurochemistry and the guests.'

'Are you really sorry?'

'No. It's just an expression. I don't feel sorry.'

'Liar. You're in love with your parasite. It lets you ride your big steel cock through space and call it 'heroic.''

'When discussing our alien counterparts, we prefer the term 'guests.''

'Fuck you and your guest.'

She stared at my face again. I was wearing dark glasses, old fashioned metal aviators with opaque gray lenses.

'I always wondered if what they said about women's biology was true or just a story to cover up some deeper, darker secret.

'What kind of secret?'

'If I knew that, I wouldn't need to go, would I? I'd still like to go. Even if I died on the way.'

'But you can't go. Maybe if you were lucky enough to be a transgender. Surgery. Hormone treatments. That sometimes works.'

'Now, you're making fun of me.'

'Am I? I don't think so.'

'If I changed myself like that I wouldn't be me anymore. Then, when I found out your secrets, I still wouldn't know.'

'There are no secrets. It's just space up there. Oceans of nothing. Mists of frozen dust. Arcs of fire and curious light.'

'I thought you were blind. Or is that another lie?'

'All cosmonauts are blind. Human eyes and optic nerves are too easily fried by cosmic rays. Giving up our eyes is the beginning of the alteration. Accepting our guests is the end.'

'Then, how do you see?'

'We don't. Our guests do.'

She leaned forward and tapped the back of my right hand, tracing my tattoos with one of her ragged fingernails.

'Stop talking about yourself in the plural, all right? Try saying 'I' occasionally. Saying 'we' and 'our' makes you sound like an ant in a big colony. You're not an ant, are you? Just a bug hooked up to an alien hive mind?'

'No. I'm just me.'

'What a relief. You should have told me you were going. You should have asked me to come with you.'

'You couldn't come with me.'

'But I didn't get to say that or think it or argue with you about it. You were just gone and then you were dead. Then I sort of died. Then you were alive again and out in space and I was here and still dead.'

'I'm sorry.'

'No, you're not, so stop saying it.'

'Would you like more coffee?'

'Do you fuck, Arkadi?'

My guest twisted uncomfortably inside me.

'What?'

'Do spacemen fuck? Do you have girlfriends or boyfriends? Do you go to space station whores? Maybe you let your parasite suck your cock out in the void, where no one can see.'

'Yes, we fuck. Some of us. But, some guests are disturbed by sex. They don't the like biochemical changes we experience, so, it's not encouraged. But it's not forbidden.'

'Do you fuck, Arkadi?'

'Sometimes. Not often.'

'Who was the last person you fucked?'

'I don't remember.'

'You're lying, aren't you?'

'Yes.'

'Because you don't want to tell me about her?'

'Yes.'

'You must have feelings for her.'

'I don't have feelings.'

'Then, why won't you tell me about fucking her?'

'It would be rude and, I believe, painful for you. Just because I'm dead doesn't mean I don't have manners.'

'So, you can still tell jokes.'

'I tell the truth. I can't help if you find that funny.'

Her phone rang. Valentina touched one of her bracelets and a shimmering blue square screen unfolded in the air above her wrist. She frowned and collapsed the phone back into the bracelet.

'Your lover,' I said.

'Love has nothing to do with it.'

'That's too bad.'

'Do you ever feel lonely? Not out there with the other corpses and aliens, but here, with people who still breathe and fuck.'

'I don't feel things like loneliness.'

'Bullshit. If you're still Arkadi, if you're still remotely human at all, you must feel lonely sometimes.'

'I'm never lonely, but I'm aware of...differences.'

'Tell me.'

'Maybe it's the way our guests see.'

'Your guest, Arkadi. Not ours. We're only talking about you.'

'I can't quite focus on individuals. We...I...see everything from a great distance, like a wide angle camera lens. Sitting here

with you now, I'm also looking at everyone else in the cafe.'

'Do all cosmonauts see like that?'

'In the void, we see a full three hundred and sixty degrees or we don't see at all. Our guests don't rely on any one sense more than another. When I've gone null and given up control to my guest, it's not necessary for me to see for weeks at a time.'

'Do you want to fuck me?'

'I don't know.'

'Of course you do. It's exactly your kind of question. Binary. Yes or no. Do you want to fuck me or don't you?'

'No. Not when you're irrational.'

'You used to love angry fucking. Fucking was how we apologized, remember?'

'Not really. After the alteration, our memories aren't entirely intact.'

'But you remember some of it.'

'Yes.'

'Good. Then you remember that we were good at it. Do you want to fuck me now?'

'No.'

'I understand. Later then. We'll fuck later.'

I took another cigarette from the pack and lit it. I took my time about it, running through the calming number patterns. I thought about leaving the cafe, but during training we'd been taught the importance of politeness. Those who no longer feel pain must respect the experience of those who are still controlled by it. I considered smiling warmly, but didn't, afraid that Valentina would think I was mocking her.

'Tell me about space,' she said. 'Was it worth abandoning me and killing yourself?'

'It's not what I thought it would be. It's not what you think it is.'

'Then what is it?'

'Space is as ordinary as this street or that hotel. Once you're over the initial shock of it, space is like anywhere else. It's life. It's ordinary. Even tedious, at times, but, like life, punctuated with moments of brilliance.'

'Such as?'

'Seeing a supernova as it happens. Our guests can see a wider spectrum than humans, so I can see the gamma ray fountains streaming from pulsars.'

'What else? Tell me.'

'Trembling blue stars being born in the Horsehead Nebula. Other intelligent races. The guests are slowly introducing us. I've met living machines that find us as strange as we find them. They can't believe that fragile meat has thrown itself out into space.'

Valentina beamed at me. 'And you went off to see all those things without me. You left me on this gray rock to go swimming through the stars and having cocktails with little green men. If you ever loved me, you would have killed me before you left.'

'Don't be stupid.'

'You could kill me now. Would you do that for me? I can't live here anymore. I hate this place. I hate it so much I can't taste food anymore. I can't even see colors. I hate it even more now, listening to you and knowing what's up there. We could go back to my place. We could fuck and then you could kill me. You'd go back to space and no one would ever know.'

'I'm leaving.'

I stood and took some bills from my pocket.

'Do this for me, Arkadi. Kill me. If I was dead, too, I could go with you.'

'It doesn't work like that.'

'I don't have to go to the deep. You have quarters on one of the space stations, right? I could live there, and then I'd be there for you when you came back.'

I dropped a pile of Euros on the table and put my hand on her shoulder, certain it was the proper thing to do. I must have been right because she rested her head on my arm.

'My quarters aren't very large,' I said. 'They'd be even smaller with two.'

'I don't care.'

'I'll never love you. I can't.'

'I don't care.'

'What would you do when I was gone?'

'I'll read. I'll putter. I studied chemistry, so I'll get a job with one of the research groups. I'll be a janitor.'

'You know that most of the station crews are like me. Altered. It's not as bad as deep space up there, but living long term on a station guarantees cancer for unaltered humans. It's easy enough to clone a heart or a lung, but one speck in your brain and you'll die.'

'Your brain has survived.'

'You don't have a guest to eat your tumors.'

'You spacemen are capable of taking care of things. I've seen documentaries. Some of you have little gardens in your rooms. Some of you even have pets. Take me with you. I don't need much. I'll be your rabbit. Give me lettuce and water and rub my ears every now and then. That's all I need.'

'I left you once and you hate me for it. Every time I leave you up there, you'll hate me even more.'

'No. I love you. Do this for me. Maybe one day one of your guests will want to taste a different kind of human host, and I'll be there and ready. Then, I can go to the stars, too.'

'Then you wouldn't be you anymore and you'd never learn our secrets.'

'I don't care.'

'How will you be my rabbit if you're dead like me?'

'We don't have to worry about that for a long time. Maybe never. Just take me with you now.'

'I didn't say it before, but, yes, I want to fuck you.'

'Take me with you.'

'You're alive. You can't live with the dead.'

'I hate the living.'

'So do I,' I said and took my hand from her shoulder. 'You're too loud and too ridiculous. How can you even think with all the noise you make? You want to know if I still have any human feelings? This is what I have: I hate you all.'

'I love you, Arkadi.'

'I know.'

I went to the door, turned and said politely, 'Goodbye, Valentina.'

She threw her coffee cup, but I was already outside. I kept my gaze forward, but in my panoramic view I could see her through the cafe window, crying and cursing at me.

The tabloids were right. I am the walking dead, a zombie veering the cosmos, sustained by an alien parasite. I'm a monster, but I can refuse to be monstrous. Closing the cafe door, saying the things I did and leaving Valentina with the living was as close to an act of love as I was ever going to get.

I went to the port and rode the great diamond tether to the jump station. There, I found the first freighter heading out beyond Jupiter.

I never went to Earth again.

Isaac Leung

FROM SCIENCE FICTION TO PORNOGRAPHY
THE CULTURAL REPRESENTATIONS OF SEX MACHINES

Different social concepts concerning the family, sexual health, potency, sexual liberalism and health epidemics have been formulated and circulated as 'sexual truths', while the understandings of body, gender, subjectivity and social relations have been legitimized within the domain of sex machines. The cultural history of sex machines has been discursively constructed by the networks of power relations in a 'multiplicity' that is constituted by 'bio-medical' knowledge and its discourses. The cultural meanings of sex machines - the contested 'reality' and 'natural' identity that are being formulated in the modern West (from the nineteenth to the twentieth century) - provide me with an insightful backdrop to further examine the cultural significance of contemporary techno-sexual inventions. In this paper, I will investigate the meanings of sex machines by studying their representations in science fiction films and pornography. Within the domain of sex machines, I will attempt an overview of the meaning of science fiction films that critically and metaphorically transgress the naturalistic and realistic notions of culture, and how the images and concepts of these films represent the normative knowledge of sexuality and technology. This project is focused on three kinds of recently invented sex machines: fucking-machines, teledildonics and sex robots. In this paper I will analyze the filmic and pornographic representations of these three types of sex machines by studying *I.K.U.*, *Fucking Machines* and *Sex Machines Cams*.

'Representations' and 'SF'

This project aims to re-articulate the meanings of sexuality and technology within the domain of sex machines by studying the 'circuit of culture' (du Gay, Hall, Janes, pp3 1997) with an emphasis on the representations and productions of sex machines. Before going into the production process of sex machines, I will conduct a textual analysis of SF films and pornography and study how sex machines are represented in different visual languages. In order to articulate the representation of sex machines in SF films and pornography, one needs to understand the meanings of 'representation' and the generic characteristics of SF and pornography. In *Representation: Cultural Representations and Signifying Practices*, Stuart Hall posits that 'language...operates as a *representational* system' and 'representation through language' is 'central to the processes by which meaning is produced' (Hall, pp.1, 1997). In Hall's term, 'language' is 'signs and symbols' that are encoded not only written or spoken words but also through sound, visual images and objects; language 'stand(s) for' something 'to enable others to 'read', decode or interpret' (Hall, pp.5, 1997). In other words, 'language' constitutes 'meaning'. Therefore, to study 'representation' is to understand the signs and symbols of the language and its signified meanings. While 'culture' is primarily 'concerned with the production and the exchange of meanings between the members of a society or group', representation underlines the 'crucial roles of the symbolic domain at the very heart of social life' (Hall, pp.3, 1997). According to these definitions, the images in SF films are the symbolic meanings generated by the language of the SF genre.

In *Science Fiction*, SF writer and analyst Adam Charles Roberts established a comprehensive literary criticism of the SF genre by positioning it within the sphere of cultural and interdisciplinary studies. Traditionally, the fundamental definition of the SF genre is a 'fantastic literature' that distinguishes a fictional world from the world in which we actually live, where the 'fictional world' portrays 'imagination' rather than 'observed reality' (Roberts, pp.1, 2006). In rejecting the delineations between the two domains in the above, Roberts problematizes the definition of SF by positing it as a 'slip-page' between 'imagination' and 'observed reality'. He introduces the concept of 'cognition' and 'estrangement' by using the famous SF critic Darko Suvin's quote: '[SF is] the presence and interaction of estrangement and cognition whose main device is an imaginative framework alternative to the author's empirical environment' (Roberts, pp.8, 2006). He explains that 'cognition' is the 'rational' and 'logical' meaning that prompts us to understand and comprehend (observed reality), while 'estrangement' is the 'alienation' and alternative aspect that is unfamiliar in our everyday life (imagination). According to this definition, the main 'device' of SF is the 'cognitive estrangement' that is relevant to our world; at the same time, it is a challenge to the 'taken-for-granted' or ordinary, everyday existence of humans (Roberts, pp.8, 2006). Therefore, the 'science' of SF is the discourse that is built upon certain 'logical principles' and 'thoughtful experiments' that constitute 'scientific', 'fact' and 'truth'; and 'fiction' is 'imagination' and 'emotion' that is understood as a 'mode of critical resistance' to modernistic rationality (Roberts, pp.9, 2006; Darian-Smith, pp133, 1996). According to Roberts, the resistance to static meaning that is informed by the 'modernistic rationality' can be seen in the portrayal of the 'objects' and the 'otherness' in SF texts. SF is a genre that emphasizes the 'object' rather than the subject, the 'other' rather than the self. The themes of SF usually focus on portrayals of the materiality and the symbolic nature of non-human characters, such as robots, aliens, and advanced technology. The depictions of 'otherness' that are based on 'logical principles' and 'critical resistance' constitute the 'continuities' and 'discontinuities' of reality. That is why Roberts posits SF as a 'symbolist genre'. He thinks SF constitutes a 'symbolic manifestation' that could be specifically referenced to the real world; it 'attempt(s) to represent the world within, reproducing it in its own terms' (Roberts, pp.9, 2006). In this way, rather than allegorically 'mapping' one thing (the fictional world) on to the other thing (reality), SF symbolically opens itself up to the possibility of transcendence and unrealized interpretation (Roberts, pp.14, 2006).

While some 'realistic' fiction also uses this kind of 'symbolic device' that encodes the effect of 'discontinuities', Roberts suggests that the specific symbolism of SF is a 'transcendental or metaphysical aura' that is referential to the materiality of 'scientific' practices. In this account, SF adopts the mode of 'realist' narrative, but one created beyond the symbolic structure of 'realist' language. It 'reconfigures symbolism for our materialist age' and 'takes metaphorical constructions literally' (Roberts, pp.15, 140, 2006). For Roberts, the 'loosing-touch' and 'strangeness' in the specific symbolism in SF is the '*representations* of the world not through *reproduction* of that world'. He thinks that to read SF is to decode the 'slippage between metaphor and literality'. Therefore SF is neither merely metaphorical nor literal, and it expresses the complexity of interpretive relations between 'poetry' and 'speculative thought' (Roberts, pp. 141, 2006). The representation of SF is thus 'interpretive relations' based on the 'poetic surplus', rather than the conventional semantic structure of metaphor utilized in other genres. SF is a 'discursive space that enables the wish-fulfillment to span this gap between metaphoric and literal' (Roberts, pp. 141, 2006); it is a paradox between what 'is' and 'is not'. The idea of 'poetic surplus' in SF is similar to Slavoj Zizek's analysis of 'the paradox of a fantasmatic element'. Zizek posits that the 'not quite human' nature of 'aliens' are 'baptized' as the 'paradoxical uncanny object that stands for what is the perceived positive'; the symbol of the 'aliens' thus 'serves as the driving force' of desire. Such a process in the perception of 'aliens' 'designates the excess over the satisfaction brought about by the positive, empirical properties of the object', and makes the 'object more than itself' creating the 'surplus-enjoyment' (*plus-de jouir*). The 'other' in SF symbolically contains both the 'empirical properties' and also the 'surplus-enjoyment', and is why Zizek uses the term 'between' in conceptualizing the relationships of 'symbolic fiction' and 'fantasmatic spectre' in the symbolism of the 'aliens' (Zizek, Butler, Stephens pp.236, 2006).

Let's go back to Stuart Hall's conceptualization of language and its meanings, where 'representation' is the study of the meaning of 'signs and symbols' underlining the 'crucial roles of the symbolic domain at the very heart of social life' (Hall, pp.3, 1997). The 'symbolic domain' of SF that is 'drained of transcendental or metaphysical aura' is referential to the materiality of 'scientific' practices, while the 'social life' is the 'alternative', 'imaginary' life of 'others'. The symbolic meanings of sexuality and technology based on the narration of the 'scientific' practices (cognitive logics) and the 'alternative' imaginary life (estrangement) of sex machines is what will be examined here. Given that the 'history' of sexuality is known as a 'science fiction'[1], specifically in the realm of sex machines, I question how the symbolic meaning of SF refers to and 'represents' the 'science fiction' of sexuality and technology in the 'reality. Based on the conceptual framework of SF in the above, what kind of symbolism does the portrayal of sexuality in SF represent?

Sexual and technological 'surplus' of sex machines in SF

In *Queer as Traitor, the Traitor as Queer: Denaturalizing Concepts of Nationhood, Species, and Sexuality*, queer and science fiction analyst Wendy Pearson states that SF is 'able to ironically replay the linkage between queerness and treachery in order to interrogate the naturalization of the concepts of nationhood, species, and in particular, sexuality' (Ketterer, pp. 77, 2004). The symbolic device of SF allows sexuality to be symbolically re-articulated in a manner that is unusual for 'natural' cultural assumptions. The symbolism of sexuality in SF is thus made meaningful by the readers according to their own wish fulfillments, and allows their imaginations to poetically play with socially assumed ideas of body, gender and sexual acts. This 'alternative' portrayal of sexuality is based on and referential to the 'continuities' and 'discontinuities' of the meaning of sexuality in the real world; it reflects the 'normal' understandings of body, gender and sexual acts, and at the same time, opens up a new space to rethink and re-explore a socially assumed sexuality.

Along with a sexual openness informed by the sexual liberation movements of the 1960s, sexual representations in SF films have also become more widely accepted. While the nature of SF depicts the cognitive logics of 'scientific' practices, the merging of sex and machinery, which underlines the relationship between bodies and machines, also became a popular topic in SF texts. Sex machines are not only being literally depicted as a new kind of body, they also represent a new kind of material and symbolic sexuality. The materiality and technologies of different types of fictional sex machines portrayed in SF films creates 'symbolic surplus' and invites audiences to re-imagine different normative notions of sexuality. While Roberts posits that the nature of SF is the 'discursive space that enables the wish-fulfillment' that lies between 'metaphoric' and 'literal', the representation of sex machines in SF is referential to the 'discursive space' of sexuality that is formulated in the real world. In many SF films, sex machines are metaphorically depicted as an alternate and imaginary world/space (of sex). For examples, George Luca's first feature length film, *THX 1138*[2] and Woody Allen's [3], which were produced in the 1970s, portray a dystopian space where sexuality is controlled by governmental systems by way of technology. Instead

[1] In The Science/Fiction of Sex: Feminist Deconstruction and the Vocabularies of Heterosex, based on Foucault's discursive analysis of the history of Western sexuality, critical feminist Annie Potts posits that the ideas and experiences of the body, sex and gender are science fictionalized by bio-medical discourses.

[2] THX 1138 is a 1971 SF film directed by George Lucas. The film is set in the 25th century where thousands of nameless shaven-headed citizens are working in a huge underground nation where sexual desire, love affairs and emotions are forbidden. The protagonist, THX 1138 played by Robert Duvall, works in a dangerous nuclear factory in a society where hard work, increased production, the prevention of accidents and the consumption of products are the only values respected by the state. His emotions are regulated by government prescribed drugs; his libido is satisfied by a mechanical masturbator and senseless holographic programs. Concerned about his problems, he goes to one of the confession booths that are placed around the city like our phone booths. He talks to a picture of a Jesus-like man called Omm, who responds in a repetitive pre-recorded voice saying things like 'Buy more, buy more, buy and be happy'. However, THX 1138, unlike other citizens in the underground state, is not happy with his life. THX 1138's assigned roommate, LUH 3417, who is also uncomfortable in this totalitarian regime stops taking her state prescribed drugs and purposely switches THX's usual sedatives with stimulants. They both begin to experience authentic emotions and happiness. They fall in love and engage in sexual intercourse, which eventually leads to them being arrested and charged with drug evasion, malicious sexual perversion and transgression (sexual activities excepting masturbation is considered illegal). At the end of the film, THX 1138 successfully escapes from the authorities to the outerworld.

[3] Sleeper is a 1973 SF comedy directed by Woody Allen. The story is situated in 2173 after a global nuclear war when the United States of America is ruled by a totalitarian leader (who looks like the Pope in wheelchair). The protagonist Miles Monroe, played by Woody Allen, has been cryogenically frozen for 200 years. He is revived in the year of 2173, and soon becomes unwillingly involved with the anti-government underground due to his advantage of being the only member in this society without a known biometric identification. As in THX 1138, the police state dictates everything from food to sexual activity. Miles is arrested and brainwashed by robot cops and scientists due to his illegal citizenship and pre-dystopian intellectual spirit. Since everyone in the country is programmed to be either frigid or impotent, an elevator-like electromechanical device called Orgasmatron is being used to induce orgasm. Orgasmatron can accommodate multiple partners, immediately giving the users sexual orgasms without the need for bodily contact.

Figure 1: THX 1138

of utilizing flashy high-tech special effects, both films present an abstract, modernist and minimal kind of aesthetic. The iconic 'white on white' *mise-en-scène* and clinical aesthetics where human warmth is absent negatively represent the 'bio' technological future (Figure 1). Situated in an alternate world of sterile white walls, corridors and rooms where windows, doors and sky are absent, *THX 1138* aesthetically and conceptually challenges the audience's 'ordinary awareness of size, dimension and perspective' in relation to actual societies (Sobchack, pp97, 1997). It comments on consumer culture, the medical industry and religious control that are spied upon by the advanced technology. In this totalitarian state, the protagonist THX 1130 is portrayed as a character that resists the programmed normative customs. He is the 'other' who fights against the state, someone who wants to escape to the outer world. Apart from the over-sanitized settings, the fictional mechanical masturbator and senseless holographic programs are used to symbolize the forbidden (human/human) sexual contact. The government prescribed sedative medicine is also depicted as a form of 'sexual machinery/ technology' that regulates sexual desire, love affairs and attendant emotions. Similarly

Figure 2: The "Orgasmatron" in Sleeper

in *Sleeper*, a fictional sex machine that looks like an elevator, called The 'Orgasmatron' (Figure 2), is used to induce sexual orgasms without the need for bodily contact. This device symbolically represents the bio-political control of the totalitarian state where all the citizens are programmed to be frigid and impotent. The protagonists of both films are depicted as the 'other', individuals who are trying to escape from the normative sex of these imagined societies (non-bodily sex). The 'other' of both films are referential to, in actual reality, 'us'. The 'discursive space' of imagination that is informed by the material setting of both states in *THX 1138* and *Sleeper*, the 'white on white' *mise-en-scène*, the mechanical masturbator, the sedative medicine and the 'Orgasmatron', are symbolic reversals of the discursively formulated space of sexuality as it actually exists in the here and now. Many sex machines were invented as a product of discursive medical and technological governance; both films invite us to transcend and re-imagine the socially constructed idea of the sexual norm. The portrayal of the alternative space in both films encourages us to re-define what is to be normative or alternative, and it allows us to create a new imaginary space that could possibly transcend both domains.

While many SF films depict the notion of merging sexuality (human) and technology (machines) by portraying the new bodies of sex machines, they also formulate a 'surplus' space for sexual imagination, specifically in regards to family, gender, and sexual health. The static, natural and 'taken-for-granted' narratives of sexuality are thus radically re-arranged by methods of exaggeration and/or counteraction in the 'discursive space 'of the new wave SF texts. For example, unlike the minimalist treatment of the future world in *THX 1138* and *Sleeper*, SF films such as *Barbarella*[4] and *Flesh Gordon 2:*

[4] Directed by Roger Vadim, Barbarella is a 1968 erotic science fiction film based on a French adult comic strip of the same name. The story is situated in the 41st century, where an astronaut called Barbarella engages in futuristic tongue-in-cheek sex during her adventurous space journeys. Barbarella is requested to stop a civil war by searching for the evil scientist Durand Durand in the city of Sogo. Along the way, she teams up with strange characters at a planet called Tau Ceti who helps her on her quest. She eventually rescues the earth with the help of the Black Queen from Sogo. The film ends with Barbarella lying in the Excessive Machine that sexually fondles her. Barbarella realizes the man who made the machine is Durand Durand.

Flesh Gordon Meets the Cosmic Cheerleaders[5] invite us to re-imagine gender and sexual acts through campy takes on fictionalized sex machines. Also known as a New Wave of SF films, during the period of 1960s and 1970s, SF films are also seen as 'sextrapolation' (Pearson, Hollinger, Gordon, pp. 52, 2008) and 'sexploitation' (James, Mendlesohn, pp. 91, 2003) epics, which intentionally expand the 'boundaries of the sexual content that can be depicted in mainstream films' (Fraiser, pp6, 1997). In *Barbarella* and *Flesh Gordon 2: Flesh Gordon Meets the Cosmic Cheerleaders*, the exaggerated and 'expanded' juxtaposition of femininity and masculinity represents the sense of campiness that 'extrapolates' men/women gender characteristics. 'Campy' is understood as the dramatic and theatrical performance that is not inherent in the person or thing itself, but in 'the tension between that person or thing and the context or association' (Newton, pp107, 1979), such as the notion of expanding the non-given gender symbols as seen in *Barbarella*. The main actress Jane Fonda is portrayed as a hypersexualized female astronaut who is being transformed into a fetishized sex object. The film's highly unrealistic

mod and kitschy costumes, the colorful spaceship and the imagined outer spaces promotes her as the impossible feminized sex icon (Figure 3). Provided that a female astronaut was impossible during the 1960s, *Barbarella*, on one hand, disrupts the cultural logics of gendered science, and on the other hand, exaggerates the sexual role and aesthetics of woman to an improbable extreme. The estranged logics of gender in *Barbarella* opened up a new debate regarding gender representations in SF films during that period. Apart

Figure 3: Barbarella in the spaceship

from the fictional 'technologies' of gender construction, the sex machine of *Barbarella* is represented by sex enhancing pills and the pleasurable and yet fatal 'Excessive Machine'. Contrary to the sedative drugs and the mechanical masturbator in *THX 1138*, the 'scientific' objects in *Barbarella* symbolize the transcendent pleasure that is informed by diverse sexual activities. Even though Jane Fonda performs as a fetishized sex object (for men, and in juxtaposition to masculinity), *Barbarella* creates an imaginary space that is bounded not by physiological limits, but in a transcendent form of non-bodily cerebral pleasure through the use of sex pills. In the last scene, the 'Excessive Machine' symbolizes another kind of pleasure that is 'excessive' and powerful enough to blur the boundaries and principles between pleasure and torture. While

the sex machines in *Barbarella* symbolize a campy femininity, *Flesh Gordon 2* can be viewed as symbolizing excessive masculinity. Similar to *Barbarella*, *Flesh Gordon 2* uses theatrical, 'painterly', colorful and campy backdrops and props that can be compared to 'science' (Figure 4). The main actor Vince Murdocco is portrayed as an excessively potent and virile hero that can combat a fictional sex machine that transmits impotent-inducing radiation. The Evil Presence from an outer planet in *Flesh Gordon 2*, who is shamed by his

Figure 4: Flesh Gordon 2 and its excessive masculinity

impotency, symbolizes the contradictory notion of essentialized maleness and the natural procreative capability by repositioning male genitalia away from scientific and medical symbols. With the aid of 'camp' elements, *Barbarella* and *Flesh*

---

[5] Flesh Gordon 2: Flesh Gordon Meets the Cosmic Cheerleaders is a sequel to the pornographic cult film Flesh Gordon in 1974. The film portrays an emperor, Evil Presence, who comes from another planet. Evil Presence wants to become the only potent man in the Universe. He threatens to make every man impotent by using his sex machine that generates an impotence ray. The story starts with the protagonist Flesh Gordon being kidnapped by the cosmic cheerleaders Babs, Sushi and Candy Love. He is taken to the Ice Planet to sexually serve the cosmic cheerleader Robunda Hooters, since all the men on that planet have been affected by the impotence ray. Later on, Flesh Gordon's girlfriend Dale Ardor and Dr. Flexi Jerkoff come to rescue Flesh, where upon, Evil Presence freezes Dale and drags her to the airship. Flesh and Dr. Jerkoff begin looking for Dale, during which time they successfully destroy the impotent ray.

Gordon 2 manifest and transform the impossible cultural 'standard' of feminine and masculine stylizations of the body. They problematize the natural understanding of sexuality and gender by parodying the 'surplus' of gender and sexual conventions. The 'Excessive Machine' and the impotent-inducing radiation machine, symbolize the instability of rational ideas concerning gender and sexual experience that is represented in realistic fictions.

Apart from their radical illustration of governance, sexuality and gender, many SF films portray new post-human species and viruses that are related to the contemporary socialization of families along with public health issues. For example, in the re-imagining of the family in SF films, birth is often depicted through the lens of reproductive technologies: cloning, robot child, male pregnancy, monstrous birth and many ways of reproducing artificial offspring are re-imagined and moved away from the solidarity of family structures, such as heritage and lineage and the national identity of social members. Many classic SF novels such as *Frankenstein* and *Brave New World* encode a direct human confrontation with newborn creatures.

In the blockbuster SF film, *Artificial Intelligence: A.I.*[6], instead of a clinical or campy setting, the audience is presented with flashy computer-generated special effects. Like many of the SF films produced from the 1980s and onward, *Artificial Intelligence: A.I.* is stylized through the language of virtual reality, with a focus on computational artifice made possible by the advanced systems of CGI. The film's main location is situated in a computer-generated metropolis that is filled with hover-car highways and flashy neon lights, in contrast to the submerged underworld of a re-imagined New York City in other scenes of the film. *A.I.* narrates the story of an artificial offspring who is on a quest to find his own identity. The portrayal of David, a new cloned and robotic child, radically alters the notion of humanity and family. The fear towards David as the 'other', a machine that indicates a disruption in the harmonious organic unity of family, challenges the audience to re-imagine the binary boundaries of machine and organism and the meaning of what a legitimate family is or can be. The confusion (love and fear) towards the David and the questions of whether he fits into established notions of family problematizes the natural heritable traits of human species and the basics of biologically based socializing units. While the David character in *A.I.* is understood as a new kind of species, the re-imagination of social units in SF films is not always confined to new fictional species. The socialization of different citizens is also defined by microorganisms, such as viruses, that prompt us to re-think concepts of public health risk management. Also situated in a computer-generated city, *Demolition Man*[7] presents an imagined combination of existing big cities in California. In *Demolition Man*, a sex machine

---

[6] Artificial Intelligence: A.I. is a 2001 science fiction film written and directed by Steven Spielberg. The story concerns a future world sometime after a global warming ecological disaster. Scientists create androids to maintain civilization since there has been a huge reduction in population. Due to the exhaustion of natural resources on earth, only licensed couples can have children. Human reproduction is generally prohibited. David, the main character of A.I., is a robotic boy with biological appearance created by Professor Hobby by a private firm called Cybertronic Manufacturing. He is a new form of robot known as 'Mecha' who is programmed with the capability to love and dream. One of the workers of Cybertronic Manufacturing, Henry Swinton and his wife Monica have decided to adopt David since their son Martin has been cryogenically frozen due to chronic illness. As David continues to live with the Swintons, he starts to feel love for Monica due to his activated imprinting protocol. But things go wrong when the biological son Martin is miraculously cured, leading to the android son David being abandoned. Just like the story in Pinocchio, David tries to look for ways to become a 'real' boy. He's convinced that Monica will love him and take him back if he can find the Blue Fairy. 2000 years later, long after the human extinction, the future alien-looking robots allow David to reunite with his mother, and this is when David can finally reach 'a place where dreams are born'.

[7] Demolition Man is a film inspired by Woody Allen's Sleeper in many aspects. The story is situated in the future world of 2032. The city San Angeles portrayed in the film (created from the joining of Los Angeles, Santa Barbara and San Diego after a massive earthquake in 2010) is a sanitized city ruled by the fascist leader, Dr. Raymond Cocteau. Due to the heavy crime, the spread of disease and the decay of natural environment in America during the 21st century, the new city prohibits anything 'bad for you'. There are laws against smoking, drinking, fighting and sexual contact in the futuristic city of San Angeles. The story begins when the criminal Simon Phoenix and the police officer John Spartan are released from a cryogenic prison after 36 years. Simon Phoenix is hired by the fascist leader to destroy the subversive underground, while John Spartan is teamed up with the innocent cops Huxley and Garcia to revolt against the authority. At the end of the film San Angeles returns to being a city with individual choice and freedom.
The rationale of banning unhealthy food, tobaccos, alcohol, caffeine and bodily contact in San Angeles is aimed at facilitating public health. In Demolition Man, sexual diseases transmitted through physical contacts are regarded as the major cause of the society downfall of the 21st century. When John Spartan and Huxley are attempting to have sex, Huxley says, 'After AIDS, there was NRS, then there was UBT'. Huxley is disgusted by Spartan's request of sex with bodily contact since a new form of sex called 'Vir-Sex' in prevalent in 2030. In that future city, people wear a new kind of sex machine on their head during 'Vir-Sex' to replace physical intercourse.

induces sexual simulation to avoid bodily contracted STDs. By exaggerating the social fears of risk and hazard that could be contracted through bodily sex, this machine symbolizes the failure of public health governance that is based on the social and medical understanding of STDs viruses in the future world. *Demolition Man* invites the audience to re-think the meaning of virus and epidemic that is referential to the AIDS epidemiology in the real world. The machine serves as a symbolic object of epidemic governance; it invites us to re-think the relationship between sexual body and viruses and the social politics behind them. Both *A.I.* and *Demolition Man* create a new domain of scientific imagination by introducing a new 'machinery' of species and viruses in contrast to the normative social narration of family and public health. The 'surplus' symbols of these new species and viruses is played out as a paradox in which David in *A.I.* is neither an insider nor the outsider in the family, and bodily sex in *Demolition Man* is neither risky nor non-risky. The fictional depictions of sex machines in both films create unease towards the accepted understanding of socialization regarding family and sexually transmitted epidemics.

The design and visualization of different sex machines in SF films are, on one hand, cognized by the materiality of 'scientific' logics of sex machines; on the other hand, they create a 'transcendental or metaphysical aura' that symbolizes the otherness and imaginary sexual life offered by the futuristic sexual technologies. The alternative meanings of sexuality such as family, health, gender, liberalism and epidemic are metaphorically exaggerated and/or counteracted in reference to the real world. The contested meanings and logics of sexuality in different SF films poetically create a surplus and paradoxical space making re-imaginations possible. While the sex machines are metaphorically symbolized but literally illustrated in SF films, they are also representing a slippage between the 'literal' and 'metaphorical' where new identities are being re-formulated. The notion of 'slippage' in SF is a logical extension of the sub-genre of cyberpunk, which depicts the 'virtual realities'. In the contemporary world, the invention of fucking-machines, teledildonics and sex robots are all made possible with the emergence of cyberspace. I will now analyze the sub-generic characteristics of cyberpunk in relation to *I.K.U.*, *Fucking Machines* and *Sex Machines Cams*.

The representations of fucking-machines, teledildonics and sex robots

'This is a figural process presently best observed in a whole mode of contemporary entertainment literature, which one is tempted to characterize as 'high tech paranoia', in which the circuits and networks of some putative global computer hookup are narratively mobilized by labyrinthine conspiracies of autonomous but deadly interlocking and competing information agencies in a complexity often beyond the capacity of the normal reading mind.' (Frederic Jameson, pp. 38, 1992)

In the above, I analyzed different examples of SF films that depict and emphasize the material nature of the sex machine. Provided that the world of technologies is not only bound by its physical existence, but also by its untouchable networks, many SF films portray an intermediated culture; these are categorized under the genre called cyberpunk. The term cyberpunk is drawn from cybernetics, which is an interdisciplinary study of a new science that combines the communication theory and control theory that encompass the 'human mind', the 'human body' and the 'world of automatic machines'. Rather than focusing on the materiality of the engineered body, cybernetics concerns the 'communication network based upon the accurate reproduction and exchange of signals in time and space' and the communication between organisms and machines (Featherstone, Burrows, pp. 2, 1995). While the term cyberpunk specifically refers to the sub-genre of SF that was built around the work of William Gibson and other writers who depicted a future world of cyberspaces within 'technological development and power struggles', the term also theoretically reconstructs the 'social theory of the present and near future' and proposes 'experimental lifestyles and subcultures' (Featherstone, Burrows, pp. 3, 1995). The symbolism within the cyberpunk genre expresses a transcendent and compressed 'time and space' communication network based on the empirical experiences of human/machine interconnectedness. It is a special site with a lack of distinct 'situatedness', with 'no fixed geographic referent in the physical landscape', and what Gibson posits as the 'non-place' (Sabin, pp 62, 1999). 'Non-space', a term for a space doesn't exist, is also a metaphor for the 'expression of transnational corporate realities as it is of global paranoia itself'. It is an 'exceptional literary realization within a predominantly visual aural postmodern production' (Jameson, pp. 38, 1992). In contrary to the 'alternate' space that I analyzed for the New Wave SF, cyberpunk's spatial narrative is the 'outright representation of the present'. It turns the fictional language of SF into a 'mere 'realism' ' (Jameson, pp. 286, 1992). This mode of speculative fiction goes 'beyond that with the image surrogate themselves (simulations, 'construct', holograms)' that is 'extended' from the physical and visceral contact between the mechanical and biological in the 'traditional fictional living space' (Slusser, Shippey, pp. 3, 1992). Given that in the earlier analysis of SF, with symbols that resisted static meanings informed by the 'modernistic rationality', SF is a genre distinguished from many 'realist' texts with an emphasis on the 'object' rather than the subject. The cyberpunk genre is seen to be less resistant to the 'modernistic rationality', but rather embraces the postmodernity of the contested meaning of culture. Frederic Jameson states 'there is an exchange and a dialectical multiplication of imaginary entities between subject and object' and 'languages through representation' in itself is 'pressed to some absolute limit' (Jameson, pp. 137, 1992).

The fantastic 'non-space' of I.K.U.

Blade Runner, directed by Ridley Scott in 1982, marked a shift in visualizing the cyberpunk sensibilities in popular and cinematic contexts. By depicting the romanticism and paranoia towards the manufactured sex android that is known as the 'pleasure model', *Blade Runner* symbolizes the cyber-sexual embodiment and disembodiment that is possible through the *mise-en-scène* of computers, televisions and billboards that broadcast the simulated organism and cloning pleasure. Similarly, self-stated as a symbolic extension of *Blade Runner*'s sexual narratives, *I.K.U.* invites the audiences to 'imagine a post-*Blade Runner*, post-apocalyptic metropolis, populated with irresistible kinky cyborgs known as 'I.K.U. Coders' (cityofwomen.org). Despite this narrative sequel, *Blade Runner* depicts Los Angeles in the year 2019 where 'replicants' are being genetically manufactured by Tyrell Corp. *I.K.U.* imagines a future Tokyo in the year 2030 where the Genom Corporation, the worldwide leader in the field of pornography, invents new sexual technologies including sex robots, orgasm data storage and sexual-presence devices. The film starts off by introducing Reiko, a type of bioengineered robot also known as 'I.K.U. Coder' (an equivalent of the 'replicants' in *Blade Runner*) that is cloned to have the ability to collect sexual orgasm data. After being activated by having sexual intercourse with a female-to-male transsexual supervisor, 'I.K.U. Runner', these robots travel around the world in teams to collect information on different varieties of pleasurable experience. During their quest for orgasm data, the film includes scenes of pornographic action between the Reiko and all sorts of partners with different genders and sexual interests. Reiko's arm can transform itself into a penis-shaped device, which allows large amounts of erotic data to be transmitted and collected during the climax of penetrative sex. Subsequently, the 'I.K.U. Runner' is sent by the Genom Corporation to collect orgasm data by inserting a machine called 'Dildo Gun' into Reiko's vagina. Instead of depicting the specialist police called 'blade runners' who are trained to destroy the 'replicants', Cheang, later in the film, introduces the counterforce through 'Tokyo Rose', a different kind of robot dispatched by another competitive I.T. company, Bio Link Corporation, in order to steal and damage Reiko's orgasm data by spreading a sexual virus. By the end of the film, an artificial orgasm system is successfully built by the protagonists and Genom Corporation begins to sell the 'I.K.U. Chips' all over the world via vending machines, while a portable videophone called 'Net

Glass Phone' becomes popular for clients in order to decode, through the broad band internet, the orgasm data of 'I.K.U. Chips'.

The cultural context of the 'Tokyo'

The spatial depiction of *I.K.U.* is the communication network that is situated in an imaginary Tokyo in 2030. Unlike the mainstream cyberpunk films such as *Blade Runner*, *I.K.U.* does not depict the material landscape of high-rise Tokyo metropolis. Cheang envisions the future world by swirling between existence and non-existence throughout the film. *I.K.U.* either utilizes existing transportation structures such as elevators, tunnels, highways and car park, or features a psychedelic and phantasmagorical animation special effect that is equivalent to a LSD induced hallucination. While the 'real' Tokyo is not actually shown, the spatiality of *I.K.U.* is metaphorically translated as a 'passage' that is a 'four-dimensional space-time manifold' (Sim, pp.18, 2001; Merrell, pp. 151, 1995). Unlike many of the films that I mentioned earlier, this 'passage' in *I.K.U.* is a process of transition; without indicating the landscape of the future 'Tokyo' as either dystopic or utopic, Cheang narrates a neutral 'non-space' that is neither pleasant or unpleasant. It is a space that is an 'absence of both existence and non-existence' (Pirie, pp. 623, 1858) and it is a 'concept of the moment' that can be 'imagined as occurring in a multi-dimensional matrix, a spatiotemporal convergence of discourses of difference and identity'. In other words, the 'passage' in *I.K.U.* encodes different contested and fragmented meanings and is a space full of wonders. While the 'Tokyo' in *I.K.U.* symbolizes a 'non-space', this space is referential and seen as an extension of the 'real' Tokyo that is embraced by global capitalism that allows the transnational 'flow' of information. Unlike many cyberpunk films that portray non-sexual corporations, such as Coca-cola in *Blade Runner* and KFC in *Demolition Man*, *I.K.U.* portrays a transnational corporation that provides telecommunications and information for the processing of sex and pornography.

At the beginning of the film, Cheang shows a TV commercial of the Genom Corporation for its cyber sex products. The advertisement is stylized by scientific and medical illustrations that are comically animated like those in the Japanese anime and manga (Figure 5). The complicated mechanisms of the 'I.K.U. system' is thus storyboarded by stylistic symbols that serve as a façade of the information (sexual) technology in which no one seems to have the need to fully understand. Being the first scene of *I.K.U.*, the TV commercial of the Genom Corporation symbolically introduces the notion of cyberpunk; the commercial represents a transnational corporation that is leading a global industrial integration of sexual sensibilities without the 'trace of material inertia'. Just as Bill Gates described the cyberspace as the 'friction-free capitalism', *I.K.U.* posits sexuality into the logic of 'friction-free' embraced by the sexual technology within the notions of global capitalism (vending machines selling orgasm data worldwide). It is a sphere where individuals 'self-satisfy' their 'erotic imaginations' and 'social fantasy' under the 'frictionless flow of images and messages' (Zizek, pp. 156, 1997). The scientific and yet comical TV commercial in the first scene of *I.K.U.* creates a space for the spectators to imagine the fictional products that is distributed by the future global flow of capitalism. Apart from the TV commercial that depicts scientific and yet comical images, throughout the whole film, the 'flow' of images and messages are represented by a mixture of traditional recorded cinema and digital effects. For example, scenes in which Reiko is having sexual intercourses with different charac-

ters are superimposed with 3D animations, X-rays, speeding and deformed images, and digitized letters and codes (Figure 6). The recorded 'real' images of sexual act are coherently and incoherently mapped, mutated and eliminated by the 'animated' scenes. While the

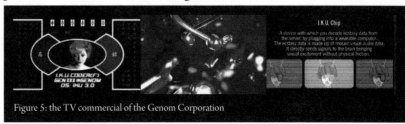

Figure 5: the TV commercial of the Genom Corporation

language of photography and digital imagination becomes fluid, *I.K.U.* metaphorically symbolizes the 'realness' of sex that can be 'animated' by free-floating artificial images.

The visualization of simulated sex is especially complex in the scene of the 'cartoon box house' which it depicts Reiko

Figure 6: sexual intercourses superimposed by 3D animations

#5, Aso Miyu, trying to acquire the orgasm data from a human, Gogota, inside a cardboard box house that is under a high-speed highway. The façade of the house is recorded by a camera which depicts the three dimensional world. Though only going once into the interior, you immediately get a sense of the virtual world's cyber aesthetics. Starting with an image of a fully-naked Gogota penetrating a blow-up doll next to a laptop and webcam, the scene goes on to Reiko joining the orgy with five other blow-up dolls floating around within the space (Figure 7). All the walls inside the house are covered by abstract animations of shapes and texts; they visually transport the viewer to another unidentifiable time and space. Watching the interior of the house is similar to the experience of going into the virtual world of network-mediated video games, where the fantasy avatars role-play through multiple identities in a graphical environment virtually inhabited by many players from all over the world. This scene in *I.K.U.* intensifies the hallucination effect by having numerous of floating squares which are the reflection of the orgy's body parts. The film's viewers are unable to identify who is who and who is having sex with which body. The last part of this scene portrays Reiko's arm turning into a penis-shape device which is inserted into Gogota's anus, and goes on to a LED-like display with patterns of coded mosaics that indicates that Reiko has successfully acquired Gogota's orgasm data. The complex and hallucinated visual treatment of the

Figure 7: "cartoon box house"

'cartoon box house' implies the visual rhetoric of postmodernity; it suggests the dislocation and disembodiment of the cybersex culture, a space that is not bounded by geographical locations and a body that is transformed beyond physicality. While the space in *I.K.U.* is non-geographical, the body of Reiko (cloned robot) and Gogota (human) are fluidly identical as they 'can take on almost limitless embodiment and have no terrestrial weight' (Welton, pp.103, 1998). Their minds are symbolically detached from the fleshy material 'weight' while their sexual pleasures and identities are re-configured into the domain of the abstract information patterns of cyberspace. The 'cartoon box house' scene's numerous floating rectangles reflecting the body parts of Reiko and Gogota symbolizes the fragmented body. Rather than symbolically illustrating a coherent mirror of the body in the cyberspace, the disembodiment of the body goes beyond the 'mirror stage'. The bodies of Reiko and Gogota are seen to be de-unified as in the partial reflections. In this incoherent imagery, the body is thus 'caught in the web of the symbolic order' (Zizek, pp.296, 2000) and it constitutes the 'armour of an alienating identity' (Fraser, Greco, pp. 173, 2005). On the one hand, the sexual 'non-space' of *I.K.U.* is visually represented by the languages of recorded cinematography and the logics of global sexual corporations. On the other hand, this space is made imaginary by the complex special effects denoting dislocation and disembodiment. By using the fusion of live recorded cinematography and animated manipulations, Cheang developed a new vocabulary of seeing and constructing the 'alienating' identities of the protagonists. In *I-k-u.com*, Cheang says that she wants to explore how 'the boundary of the human race is becoming unclear' when 'the words like 'Human Genome', 'Clone', or 'Human Robot' are being whispered'. I will analyze the 'alienating' identities of the new human race that are embraced by the 'non-space' in *I.K.U.* below.

The queerness of identity narrations in I.K.U.

Provided that Cheang wants to 'lavish visual metaphor for the sexual freedoms afforded by the internet, fantasies you can indulge with others regardless of gender, social constraints or even physical possibilities' (cityofwomen.org), *I.K.U.* purposely introduces as many sexes and sexualities as possible: biological male, biological female, FTM, drag queen, androids, orgies, one-on-one, kinky versus vanilla. Even the abstract orgasm data is symbolically rendered into an image of fragmented sexual identity. On the one hand, Reiko is portrayed as a sexualized object and commodity that is produced by the big sex corporation; Reiko's image and sexuality fit into the conditions of advanced capitalism where objects are fetishistically displaced and fantasized for (sexual) consumptions (Sedgwick, pp. 97, 1997). On the other hand, rather than positioning the objects (sex robots) as 'others' like the 'replicants' in the 'off-world colonies' of *Blade Runner*, which are classified, racialized and gendered in contrast to the 'Blade Runners' on the 'Earth', Cheang purposely 'de-problematizes' the popular dystopic visions of cyberpunk that emphasize a conflict between human/machine, male/female, hetero/homo, by using an avant-garde narrative film language. The vision of *I.K.U.* is neither dystopic nor utopic. The identities of the characters portrayed in *I.K.U.* are never made clear. The protagonists have fluid and mutable identities and they commit to their contradictory and partial nature of being within cyberspace. Spectators of *I.K.U.* can hardly distinguish or make sense of the narratives of the character's identities due to the abstract and experimental treatments of the disrupted time/space sequences. Rather than utilizing a linear logic, *I.K.U.*'s story is arbitrarily sequenced with half-perceived flickers of full-frontal views of intercourse. The visual narrative is choppy enough so that spectators, without reading the synopsis, cannot easily recognize how scenes and characters are being developed.

Concerning the fluid sexuality and identity of the characters, Cheang portrays Reiko as 'biologically' defined female, at the same time, she can perform the 'male feature' when her hands are transformed into a penis-shaped devices. In the first scenes of the film, when Reiko is being activated for her sexual function, the camera deliberately focuses on the close-up shots of the bulgy crotch of 'I.K.U. Runner' clad in underpants (Figure 8). During 'foreplay', the body of 'I.K.U. Runner' is being 'worshipped' by Reiko. 'I.K.U. Runner' is perceived as, and seems to be, a masculine African man. This scene represents the mainstream hetero erotic qualities in Japanese pornography that are seemingly 'predictable'. Though by the end of the film, when Reiko has acquired all the orgasm data, the film again reveals the close-up of the crotch of 'I.K.U.

Runner'', this time without underpants (Figure 9). Cheang unexpectedly reveals the biological gender of 'I.K.U. Runner' as female. Instead of penetrating with a biological penis, 'I.K.U. Runner' uses a 'Dildo Gun' to satisfy Reiko. *I.K.U.* disrupts the gender expectations bounded by the 'interior essence'. In *Gender Troubles*, Judith Butler once wrote, '… expectation concerning gender, that it operates as an interior essence that might be disclosed, an expectation that ends up producing the very phenomenon that it anticipates' (Butler, pp. 94, 2004). The narrative concerning the gender of 'I.K.U. Runner' rejects the logical expectations of the inside

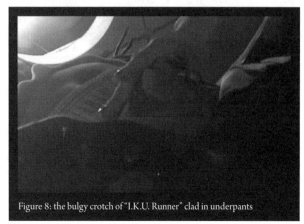

Figure 8: the bulgy crotch of "I.K.U. Runner" clad in underpants

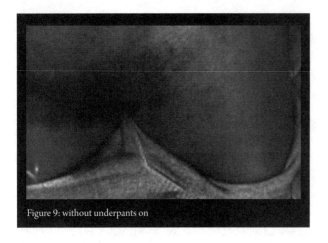
Figure 9: without underpants on

(biological gender) /outside (gender acts and gestures) gender coherence and it revolves around the 'metalepsis' of '(trans)gender performativity' which disrupts and contradicts the spectator's logical 'anticipation' regarding gender. The careful arrangements of close-ups at the beginning and the end of the film encode the fragmented notions of gender, which is mutable, partial and temporal.

The 'Dildo Gun' further renders and reduces the fluid sexual identities into pure, 'clonable' and reproducible genetic codes. All the sex acts; complex identities and interrelations between human and machine are reduced into codes of data. The 'nature' of sexual identities is thus manipulated into a 'technical product', the 'natural reality' itself becomes something 'simulated', and the only 'Real is the underlying structure of DNA' (Zizek, pp. 133, 1997) where sex is translated as the flow of data. This schema of 'reproduction' via the 'transference of genetic material' (Roof, pp. 172, 1996) symbolizes the disintegrating notion of a separated 'human or machine' identity in *I.K.U.* The illusions created by the similar costuming, make-up, gestures, and special effects of different characters make the bio-engineered robots (machines) indistinguishable from human. In the DVD version of *I.K.U.*, after the scenes showing that the successful mission of building up the *I.K.U.* system in which 'I.K.U. Chips' are sold all over the world via vending machines, viewers can choose two different abstract endings. The 'Ending Type 1' depicts Reiko meeting up with the 'I.K.U. Runner' again; despite the mission being set up by the Genom Corporation, Reiko and 'I.K.U. Runner' fall in love in a human/machine relationship and drive away on a highway. This ending tries to convey the idea that the bio-engineered robots are not the 'objects' (serving the human), they are in fact symbolically identical to the subjects (human) as the same 'species' in a future world. The 'Ending Type 2' depicts one of Reiko's human partners, 'Akira the Hustler', falling in love with the 'I.K.U. Runner'. This ending allows the viewer to consider whether 'Akira the Hustler' is a human or a machine, and correspondingly, the boundaries of his identity as a 'hustler'. The suspended and non-conclusive endings, along with fragments of dialogues, visual effects and performances further suggest an indefinite reading of the character's identities. The signs and symbols of the characters are made complex, paradoxical and impossible to be logically decoded; the 'reality' of the story is effaced, leaving all behind in an uncanny mode of interpretation and imagination within the sphere of sex and technology.

As an experimental form of cyberpunk, *I.K.U.* challenges the spectator's expectations regarding the genre and modes of traditional narrative. The textual information provided by *I.K.U.* disrupts the logical anticipation on how certain stories, identities and scenario are usually developed. The rejection of either/or narrative devices in *I.K.U.* opens up alternative possibilities for trans-racial and trans-sexual imaginings of the body. This hybrid narration is central to the genre of *I.K.U.*, one that rejects cinematic conventions by blending cyberpunk and pornography. Opposite to the representations of sex robots in *I.K.U.*, fucking-machines and teledildonics have mainly been visualized by the pornographic industry. I will discuss the pornographic industry's representation of sex machines below.

The representation of fucking-machines and teledildonics in pornography within the virtual space – from the presentability to imaginability

In *I.K.U.*, Cheang portrays a 'non-space' that is formulated by communication technology, conversely, independently made fucking-machines and teledildonics are pornographically utilized and represented in cyber space. To study the pornographic representations of sex machines, one needs to understand the symbolic meanings of pornography. Despite the very fundamental definition of pornography that is aimed at inducing sexual arousal for the consumer by portraying sexually explicit content, its cultural meaning has also been oppositionally defined by 'social and legal conflicts' that are motivated by the 'effects of mass culture, consumerism and the commodification of goods (among such, the female body) upon 'high' culture and its discourse' (Lucamante, pp. 99, 2001). While the term 'erotica' is symbolically positioned higher as 'literary or artistic items having an erotic theme', 'pornography' is understood as 'obscene', that is, characterized as 'offensive to morality or decency' and 'causing uncontrolled sexual desire' (Foerstel, pp. 185, 1997). In *Porn Studies*, rather than positioning pornography as 'obscene', film and pornography critic Linda Williams introduces a contradictory term '*on/scenity*' to illustrate the nature of pornography. She says, 'If obscenity is the term given to those sexually explicit acts that once seemed unspeakable, and were thus permanently kept off-scene, *on/scenity* is the more conflicted term with which we can mark the tension between the speakable and unspeakable which animates so many of our contemporary discourses of sexuality' (Williams, pp. 4, 2004). Provided that the porn industry in the United States makes a total of 10 to 14 billion dollars annually, she conceptualizes pornography by using Frank Rich's analysis that pornography is a cultural 'main event' that 'interrupt the more important narrative mission of film' and serves as 'an important insight…to understand the choreography of performing and laboring bodies' (Williams, pp. 5, 2004). Neither situates pornography on a moral low ground, nor on an artistic high ground; Williams notes that pornography is a filmic fantasy, 'a filmic of excess, specifically bodily excess' (McGowan, pp. 28, 2007). In *Film Bodies: Gender, Genre, and Excess*, Williams posits horror, melodrama and pornography as 'body genre', in which are encoded the 'excess' that 'catches the human (spectator's) body at its most heightened and aroused state'(Williams, pp. 333, 1999; Jermyn, Redmond, pp. 121, 2003, Grant, pp. 141, 2003). This heightened state is the surplus of the 'Real' that 'escapes its organization by the Imaginary', it renders the speakable and unspeakable in a 'frenzy of the visible', by having a 'direct access' and 'maximum exposure' on the 'opening and closing' of body parts and bodily fluids (ears, nose, rectum, mouth, penis and vagina, blood, tears, semen, etc). Through the vision of intense and excessive stimulation on screens, the (spectator's) body is affected by the 'psychic experience of the Real' and 'ecstasies' (Jagodzinski, pp. 54, 2004, Boer, pp. 61, 1999). This way, the body 'besides itself' (as in the material stimulations) has an excess value, which is the 'pleasure' in porn (orgasm), 'screams of fear' (anxiety) in horror and 'sobs of anguish' in melodrama (tears). In pornography, the spectator's bodily feelings are 'consciously and unconsciously externalized through the Other… from this Symbolic Order… and holds my body hostage to a particular 'reality''. Since the characters are 'doing' everything in excess for the spectator, it frees the spectator from involvement with, and responsibility towards, such extremes of bodily commitment. Spectators are thus being rendered in the symbolic universe of pleasure and pain, while 'it', the machinery of pornography, 'does, enjoys and suffers' everything for the spectators' (Jagodzinski, pp. 55-57, 2004). This 'in-between' notion of pornographic representation rejects the direct mirroring theory. Based on Lacanian psychoanalytical theory, the 'crack' in the mirroring process is how Judith Butler separates Imaginary from Symbolic registers: 'by accepting the radical divide between symbolic and imaginary…reconstitute sexually differentiated and hierarchized 'separate spheres'…by which the symbolic reiterates its power and thereby alters the structural sexism and homophobia of its sexual demands' (Butler, pp. 106, 1993). Under this schema of representation, pornography encodes the 'disembodied and embodied dialectic of spectatorship' (Campbell, pp. 165, 2005), thus the organization of hysterical imagery in pornography is seen to be a contradiction in itself. In *Fucking Machines* and *Sex Machines Cams*, the 'biological phallus' is replaced by the 'technological phallus'. What are the symbolic meanings of phallus displacement in fucking-machines' and teledildonics' pornography? How do contemporary robotic and networked technologies induce us to rethink the representations of pornography? I will analyze *Fucking Machines* and *Sex Machines Cams* below.

Fucking Machines and its 'excess' of Real

Fucking Machines is the first pornographic website that is solely dedicated to fucking-machines' sex. It was the second website launched after the umbrella company, Kink.com[8], was founded in 1997. Located at the former San Francisco Amory, Kink.com is a corporation famous for producing alternative fetish pornography. *Fucking Machines* is one of the Kink.com's most popular sites, having more than 350 online streaming videos depicting women being penetrated by thrusting machines. Apart from the videos, *Fucking Machines* also has a blog, forums and a section that displays all models of their fucking-machines, including the technical specs, photos and written descriptions. Rather than displaying the name and sexual features of the porn stars like a catalogue in many pornography website, *Fucking Machines* explains every details of their machines from 'Fuckzilla' that is able to walk, to 'The Lick-a-chick' that operates many prosthetic tongues at variable speeds. Under the same thematic narrative, Kink.com also launched *Butt Machine Boy* in 2003, which depicts the same machines interacting with men. But because of low subscription rates, *Butt Machine Boy* is no longer being updated.

On the front page of *Fucking Machines*, there's an introduction explaining the 'selling point' of its productions. It says, 'We take kinky sex to a new level… by sex toys and machines at speeds up to 350 RPM, leading to genuine orgasms… The experience of getting fucked by a machine brings girls hot orgasms, many shuddering in full body orgasmic bliss… If you like seeing women getting fucked by machines and having genuine screaming, cum dripping orgasms, *Fucking Machines* is the Website for you.' Obviously, besides the robotic fetish that is different from mainstream pornography, *Fucking Machines* emphasizes a narration of orgasm by displaying fully visible scenes of human/machine frictions and female ejaculations. Besides the images that visually demonstrate the orgasms, the orgasmic effects are also narrated by the performer's voice, accompanied by the strong and repetitive noise that is generated by the mechanical fucking-machines.

As it states in the introductory page, representing the 'genuine' is what *Fucking Machines* stands for, 'realness' is being 'factually' recorded in all the episodes. For example, the episode *Amateur Girl Fridays – Mason* begins with an interview that is conducted by a camerawoman. In this scene, the performer Mason sits on a stool in what appears to be a garage. Instead of having shelves that are filled with power tools, cables or automobile products, the room is filled with different kinds of fucking-machines. The *mise-en-scène* leads the spectator to realize that it is the 'storage room' of the porn studio (Figure 10). With a voice-over, the camerawoman begins to ask the Mason some basic questions regarding her personal life, and then she says, 'I'm going to make sure that you're comfortable, and I'm going to make sure that you know how things work. I'm going to leave you alone with the devices. I don't want you to fake it. You don't need to look at the camera. Anything you have seen in porn, just forget about it. I just want you to be yourself today. If any machines is not working, tell me and I will change it' (fuckingmachines.com). Right after this introduction, Mason strips and turns on the machine next to her. She starts to experiment with the machine's knobs and figures out the effects of different speeds.

[8] A quote from 'About Us' at Kink.com: 'Kink.com was started in 1997 by bondage enthusiast, Peter, who was a PhD student. After realizing consensual BDSM games were more exciting than finance, he left academia to devote his life to subjecting beautiful, willing women to strict bondage. The result was Hogtied.com, Kink's first site. Hogtied now has an enormous archives of videos depicting many tightly bound women.

Through adhering to our core values, kink.com has grown into a respected company which has attracted talented employees. Kink's team of more than 90 people is now dedicated to bringing you the most imaginative fetish material. Each of our unique websites is directed by a webmaster that is heavily kinky. Each webmaster's passion is to bring their kink to life to deliver authentic fetish footage. Our models are never told to act or artificially struggle.

BDSM is about respect and trust. When you watch a Kink.com movie, you are watching real BDSM-loving people play in this context. We at Kink.com pride ourselves in the authentic reproduction of fetish activities enjoyed by those in the BDSM lifestyle.'

The camera then moves to a close-up of the machine engine, as the thrusting sound gets stronger and stronger. After that, Mason starts to interact with the machine in different positions with a 'wish' to attain 'fulfillment'. The emphasis on the 'realistic' and 'autonomous' aspect of *Fucking Machines* on the one hand articulates the male-oriented pleasure to female pleasure; on the other hand, it reinforces the boundaries of a commodity economy that Mason is unavoidably trapped within: the 'mechanic enslavement' of the 'libidinal economy'. Mason is symbolically positioned as a 'workstation' (for men) (Pettman, pp. 123, 2006). The 'genuine' pleasure of the performers in *Fucking Machines* is contradictory and difficult to decode. Williams states in *Hard Core*,

Figure 10: "storage room" of the porn studio

'this woman is simultaneously insatiable and satisfied, capable both of continuing her pleasure indefinitely and of satisfying herself through her own efforts at clitoral stimulation' (Williams, pp. 109, 1999). Whether Mason's 'pleasure' is genuine or not, the 'realness' and 'autonomous' depictions of *Fucking Machines* is situated in between the capable and the incapable.

In addition to depicting the 'amateur' performers learning to attain pleasure for their first time by using fucking-machines, in another episode, *Masturbating addiction - Sindee Jennings*, shows an 'expert' who can 'handle' the machines like a professional. In the descriptions, it says, 'You may remember Sindee from the *Squirt Off Olympics* where she hosed Flower Tucci and Via with her squirting pussy... Sindee is back and ready for two challenges - fast machines and huge cock... She cums a river of squirt all over the machines, the set and the crew... she challenges her pussy to take big dicks... which has her squirting all over her stomach. Try and keep up with this hot girl's orgasms!' (fuckingmachines.com). During the 50 minutes of the episode, Sindee ejaculates more than five times with five different machines. The images are centered on Sindee's vulval skin and the dildo of the machines (Figure 11). The audio is a constant mixture of human tone and noise from the engines. The backdrop and the other parts of the bodies (both machines and Sindee) are no longer a 'presence' in the sex act narrative. The framing and cropping of the partial bodies that are fused with the total sound level creates an ironic incoherent tension, and urges both Sindee and the spectator to 'reconcile' their (physical and psychological) tensions through the dramatic expulsion of the squirting scene. While the video data is streaming on the spectator's screen, Sindee's fluid is smeared on the machines, her own body, and the set. It pushes the 'functions' of the machines and Sindee to a maximum level in order to reconcile the narrative tensions. I am confused as to whether she's the subject who is controlling the machines (for her heightened orgasm) or the object that is being operated by the machines (to create a surplus value for a commodity).

Figure 11: Masturbating addiction - Sindee Jennings

While the biological partner of the performers are replaced by a mechanical object, *Fucking Machines* exaggerates the 'surplus of the Real' to the maximum extent of 'sex' and minimum of 'foreplay'. It doesn't show any scenes of hugging, fondling, kissing or oral sex. It only focuses on things that are 'essential' and 'necessary'. Without adding any 'unnecessary' narrative of sex, *Fucking Machines* aims to represent and reveal the greatest amount of 'realness' as possible. Through the clearest images of HD video recording, the best craftsmanship, the most minimalistic plots and *mise-en-scène*, and the most faultless noise that could possibly be recorded from the machines, *Fucking Machines* represents a 'technical perfectibility' (Baudrillard, pp. 52, 1997) which presents the viewer with an extreme 'realism'. On the one hand, *Fucking Machines* emphasizes 'genuine orgasms' via 'machines at speeds up to 350 RPM'; on the other hand, it visualizes the 'direct access' and 'maximum exposure' to the most artificial and unreal ways of sex possible. With the 'lack' of biological human to generate orgasm, *Fucking Machines* pushes the 'surplus' of pornography to the extreme; at the same time, it doesn't present the spectators with anticipation of how certain stories, identities and scenarios are going to be developed. The logic of erotic and pornographic representations is solely 'deduced' to the reproduction of the techniques (of machines and human). This scenario illustrates what Slavoj Zizek calls the 'paradox' or 'unpresentibility' of pornography' in which 'the congruence between the filmic narrative (the unfolding of the story) and the direct display of the sexual act is structurally impossible' (Zizek, pp. 177, 1997). The gender and body identities of the performers, the machines and the spectators are constantly being negotiated between the filmic narrative and the 'realness' of sexual expressions. This contested zone is further made complex by the interactive tele-presence technologies in which the data of sex isn't recorded, but it's displayed in a real-time configuration of the body.

Sex Machine Cams and its 'lack' of Real

Sex Machine Cams is the first pornography site specializing in interface designs that allow users to control fucking-machines in real-time via the internet. Under the umbrella of *Flirt 4 Free*, a website that is dedicated to providing real-time private shows for subscribers, Sex Machine Cams offers the spectator a chance to drive the fucking-machines and chat with the performers by using the virtual interface on the website. It states on its blog, '…sex in a computer simulated virtual reality, especially computer-mediated sexual interaction between the presences of two humans…We take the whole fucking machine experience from our studio to your home. The latest in sexual entertainment' (sexmachinecams. com). Unlike *Fucking Machines*, Sex Machine Cams is produced in the studio, and is simultaneously mediated by spectators from all over the world. Without a cameraperson, the angles and framings of Sex Machine Cams are configured by the performer with an operating system that includes multiple cameras, lighting and special effects that are networked with the live broadcasting systems[9] (Figure 12).

When you go to the front page of Sex Machine Cams, you see a video banner depicting examples of the live sex show with text floating across, such as, 'Drive a Sex Machine from your Own Home', 'Real Sex Machines…Real Orgasms'. On the right hand side of the banner, there is a box indicating which performers are currently online. In addition to the flashy banners, Sex Machine Cams has a blog and calendar with an RSS feed function, so members of the website can

[9] Sex Machine Cams is operated by the TriCaster™ system which is designed for the live digital broadcastings. Here's an excerpt from the official site of TriCaster™: 'The process of creating live, network-style television can be very costly and require massive amounts of expensive equipment and a large crew of people. TriCaster™ changes all of that. In one lightweight, portable system (small enough to fit in a backpack), you have all of the tools, including live virtual sets on select models, required to produce, live stream, broadcast, and project your show.
There is a reason that TriCaster is the standard in portable live production for major players like Fox Sports, MTV, VH1, NBA D-League and the NHL. Its small footprint makes it possible to broadcast from anywhere and TriCaster is flexible enough to allow you to deliver live productions on your own or with a team. No matter where your live broadcast plans take you; there is a NewTek TriCaster perfect for you.'

receive the latest schedules and information without visiting the website. After reading the detailed information about each performer, including their birthdays, weights, heights and body sizes, members of Sex Machine Cams can login for free 'foreplay' with the performers. By clicking on one of the performers, members will go to an interface that broadcasts a live webcam image with a chat function. While the performer is covered by lingerie/under-

Figure 12: the studio of Sex Machine Cams

wear, he/she attempts to seduce the site's membership by typing erotic missives and assuming alluring poses. After the end of the free trial, members can either choose to buy credit for a private show, or to chat with another performer. This way, the line between the 'foreplay' and 'penetration' is distinguished by the presence/ non-presence of the genitals, and it is configured by the credit/ no-credit pay system.

In the private show, members are invited to an interface that has a controller with virtual knobs. The visual images of pornography are made increasingly complex in Sex Machine Cams; besides the presence of the full exposure of the body and genitals, the sex act is also enacted by writing in the chat box. While the body of the performer is fragmented by cropping different body parts, the grammar and vocabulary of the texts are also uniquely re-structured for cyber sex. 'Emoticons'[10], acronyms, abbreviations, and different linguistic strategies are applied to develop an online relationship in the shortest time possible. Performers send flirtatious messages to seduce the site's subscribers, such as 'Hello I'm Summer…ohhh talking dirty if you want', and the consumer responds with such language as, 'I'm nude…can u zoom in your puss?'[11]. The lack of correct grammar and vocabulary, the 'poetic' depictions of sound and body movements and the incoherent logics of the dialogue expand the space for imaginative sex acts. Unlike *Fucking Machines*, which is aimed at depicting visual 'realness', Sex Machine Cams represents a visual language that is completely the opposite of real. The backdrop of Sex Machine Cams is like the virtual sets we can see on *CNN* or *ESPN*, in which they simulate a working newsroom environment. With the live digital broadcastings system, the sexual performance can be visualized in locations ranging from a studio, a stage, a presentation hall, a football stadium, a spaceship, to an abstract motion background that looks like the movie *Matrix*'s introductory animation. The performer can be moved from one place to another with a simple click on the control panel, while his/her presence can be recorded by multiple camera angles. Unlike the perfect clarity of *Fucking Machines*, that depicts the clearest and most visible images of human/machine frictions and ejaculations, the performers and the machines of Sex Machine Cams can be blurred, color adjusted, distorted, reduced, sharpened and stylized by different filters. The sound emitting from the performers and the machines can be turned up and down to create a special erotic soundscape. All the visual and audio language in Sex Machine Cams represents an ironic artificialness that is completely detached from the 'natural' world. The inconsistent time and space depictions of Sex Machine Cams through its use of constructed visual and audio languages creates a surplus value of distortion that 'constantly escalates the collapse of the real into overdetermination and self-parody' (Abbinnett, pp. 112, 2003). Instead of taking the requests from spectators literally, the performers of Sex Machine Cams guarantee displaying more than what spectators expect. If being in a porn studio is not seductive enough, she/he can ride the machines like a football player or a news

[10] Pictographs that are made by keyboard symbols. For example, the smiley face.

[11] Excerpts recorded from visiting the live-show at the Sex Machine Cams studio on 3 Oct, 2008.

anchor; if the voice is not exciting enough the system will exaggerate it. The narratives of Sex Machine Cams are always expanding through the play of the body, time and space. The aesthetic value of manipulated images in Sex Machine Cams thus goes beyond the 'original' and 'natural' object. It provides the spectators a simulated experience of the 'real'.

In *Fucking Machines*, a 50 minute episode can depict multiple orgasm shots, though in teledildonics pornography, it's not always the case. Since Sex Machine Cams runs in real-time, the sex acts cannot be recorded and edited to depict images of intense orgasm one right after another. In fact, most of the spectators only stay at the private show for a short period of time, since every minute of viewing costs from 2 to 6 US dollars. Therefore, the presence of the performers is further fragmented by these time constraints. Instead of enjoying the dramatic expulsion of the squirting scenes, the pleasure of using Sex Machine Cams for the spectators is his/her control over the virtual knobs. Apart from the visual and audio manipulations that are controlled by the performers and crews, the 'effects' of Sex Machine Cams is also co-mediated by the spectator. On the virtual panel, far away from where the studio is located, the spectator can adjust the speed of the machines to 10 different levels. While the visual and audio language of Sex Machine Cams represents the ironic artificialness that is opposite to the 'natural' real world, the virtual panel represents a 'realness' that assures coherence between the real material world and the virtual one. Even though the only material actions that the spectators make are mouse-clicking and watching the screen, the symbolic sensation is heightened beyond the material pleasure. The 'realness' of the virtual knobs is thus contradicted by the 'lack of real' presence on screen, creating an ironic tension in teledildonics sex. Instead of reconciling the tension by an ejaculation that is guaranteed to be captured by camera, Sex Machine Cams always assumes a partial aesthetic and identity. Under the strictures enacted by the pay-by-the-minute system, no one has 'full-ownership' of the sex act. Unlike *Fucking Machines* where the bodily discharge symbolize the end of the pornographic narration, the performers in *Sex Machines Cams* can never 'satisfy' him/herself fully by physical ejaculation (the performers procrastinate as long as possible). Correspondingly, the spectators can never psychologically acquire the climax and resolution. The lack of a reconcilable ending in teledildonic sex renders a temporary, partial and unsustainable sexual spectatorship and a pleasure that dramatically differs from mainstream pornography on the internet.

In *Fucking Machines*, I analyzed the contradictions of narrative representation and spectatorship by using Zizek's 'paradox' or ' 'unpresentibility' of pornography'. *Sex Machines Cams* no doubt also represents the contradictory notions of pornography, though the symbolism of *Sex Machines Cams* is rendered in a reversed position. While *Fucking Machines* depicts the visual and audio 'realness', *Sex Machines Cams* encodes the notion of computer artificiality. Correspondingly, while *Fucking Machines* symbolizes the material impossibility (of scene by scene of ejaculations), *Sex Machines Cams* assures the possibility of physically remote controlling the performer. Therefore, the modes of illusionary imagination and spectatorship of *Fucking Machines* and *Sex Machines Cams* are different, even though the performers are interacting with similar fucking-machines. The symbolic articulations of body in teledildonics pornography encodes the boundless space of networked communications; instead of projecting the 'real' identities, Sex Machine Cams returns the body's identity back to its fundamental nature of artificialness. In Zizek's account, virtual sex heralds 'the end of the virtual space of symbolization' where objects are 'transitive' and 'instantly here' (Zizek, pp. 190, 1996). He posits that the fragmented presence of pornographic images creates a hyperreal situation where all objects are 'de-realized' and radically exposes the 'myth' of 'real sex' (the act with a flesh-and-blood partner) that is inherently phantasmic (Zizek, pp. 2, 1994). According to the Lacanian thesis 'there is no such thing as a sexual relationship', in reality the 'real' body only serves as a support of the subject's phantasmic projections where no coherent, perfect and harmonious sexual coupling is fundamentally possible. Zizek re-articulates this notion by proposing that all sexes are fundamentally distorted and that virtual sex 'simply renders and manifests its underlying phantasmic structure' (Zizek, pp. 2, 1994). This mode of phantasmic imagination is constructed by the illusions that the performers and spectators are simultaneously subject and object; the frame of representation (on the screen) is neither statically 'mastered' nor 'slaved' by the subject or the object. The performers and spectators are both used to make sense of the sexual act according to the logics of the computerized networks where every action and presence are immediately realized and made operable between the mediation of activity and passivity. It is a state of paradox and confusion where the information of the 'reality' is 'too much'. The capacity and dimension of imagination is thus expanded to an immeasurable metaphysics.

The representations of sex machines – a mutational process of identities

While I attempted to constructively examine the representations of sex machines by using the methods generic analysis in the above, I found that the symbolic meanings of fucking-machines, teledildonics and sex robots are re-articulated by the concept of 'surplus'. The excessive values of these contemporary sex machines' representations underlines the contested boundaries between subject/object, femininity/masculinity, existence/non-existence, utopic/dystopic and reality/fiction. Bodies, genders and the state-of-being are being rendered and mutated by the deconstructed and fragmented treatments of languages in SF films and pornography.

In SF films, sex machines are seen as symbols of governance, family, gender and virus. The languages of SF films are excessively stylized, depicting the objects, aliens and outer spaces that serve as a 'slippage' between 'cognition' and 'estrangement'. The 'slippage' in SF is then metaphorically mutated into images of a non-existent space, called 'non-space' in the cyberpunk genre. The original 'surplus' values of SF are thus further problematized by the abstract signs and symbols of the characters, stories and *mise-en-scène* of cyberpunk films. As in the analysis of cyberpunk *I.K.U.*, the representations of sex robots that I analyzed are encoded with a paradoxical mode of interpretation and imagination; the identities of bodies, genders and the notions of reality are made increasingly complex and uncanny. The signs and symbols of sex machines representations in cyberpunk are made incoherent. Correspondingly, in pornography the original 'surplus' of signs and symbols emphasizes 'it' (what cannot be shown in the non-pornographic film) where spectatorship is the pleasure of the reiteration of 'it' 'does', 'enjoys' and 'suffer' in 'maximum exposure'. The excessive stimulation is mutated into a virtually simulated pleasure where the delineation between 'it' and the spectator is no longer easily defined. This simulated space underlines the 'non-space' that is portrayed in the cyberpunk genre. The mutable 'surplus' values in SF films and pornography are further problematized by the generic categorizations of the representations of sex machines. While *I.K.U.* is included in a hybrid genre of cyberpunk and pornography, its language of representation, such as hallucinatory aesthetics and animated special effects are also applied to the teledildonics porn industry. The visual rhetoric of the penetrating scenes in the cyberspace in *I.K.U.* is thus symbolically exchangeable with the artificial presence of the performers on the screens of *Sex Machines Cams*. The generic distinctions of SF films and pornography are further erased by the reversible codes of language in the cases that I analyzed. The representations of sex machines in SF films and pornographies are emphasizing 'surplus' values, thus the notion of 'surplus' is fluidly mutated in different modes and facets within the hybrid meanings of either filmic or pornographic representations of contemporary sex machines.

This mutable 'surplus' phenomenon of sex machines provokes me to wonder about the notions of 'representation' in relation to 'reality'. As the 'representations' of sex machines are moved into the realm of imagination, I would like to investigate in what contexts these machines are being projected and imagined as these specific 'surplus' modes. How are the designs and applications of fucking-machines, teledildonics and sex robots being produced before the process of 'representation'? Provided that this paper gave me an insight into 'surplus' value, I will further investigate in the field interviews of the productions of sex machines and try to gain insight into the imaginative processes of the producers of fucking-machines, teledildonics and sex robots. In the next paper, I will aim to bring this into the larger cultural context in which contemporary sex machines are being produced.

References

Du Gay, Paul; Hall, Stuart; Janes, Linda. Doing cultural studies: the story of the Sony Walkman. Published by SAGE, 1997

Hall, Stuart. Representation: cultural representations and signifying practices. Published by SAGE, 1997

Ketterer, David. Flashes of the fantastic: selected essays from the War of the worlds centennial : Nineteenth International Conference on the Fantastic in the Arts. Greenwood Publishing Group, 2004

Sobchack, Vivian Carol. Screening space: the American science fiction film. Rutgers University Press, 1997

Pearson, Wendy Gay; Hollinger, Veronica; Gordon, Joan. Queer universes: sexualities in science fiction. Liverpool University Press, 2008.

Frasier, David K. Russ Meyer-The Life and Films: A Biography and a Comprehensive, Illustrated and Annotated Filmography and Bibliography. McFarland, 1997.

James, Edward; Mendlesohn, Farah. The Cambridge companion to science fiction. Cambridge University Press, 2003

Zizek, Slavoj; Butler, Rex; Stephens, Scott. Interrogating the real. Continuum International Publishing Group, 2006

Featherstone, Mike; Burrows, Roger. Cyberspace/cyberbodies/cyberpunk: cultures of technological embodiment. SAGE, 1995.

Sabin, Roger. Punk rock: so what? : the cultural legacy of punk. Routledge, 1999.

Slusser, George Edgar; Shippey, T. A. Fiction 2000: cyberpunk and the future of narrative. University of Georgia Press, 1992.

Jameson, Fredric. Postmodernism, Or, The Cultural Logic of Late Capitalism. Duke University Press, 1992.

Sim, Stuart. The Routledge companion to postmodernism. Routledge, 2005.

Merrell, Floyd. Semiosis in the postmodern age. Purdue University Press, 1995.

Pirie, William Robinson. An Inquiry Into the Constitution, Powers, and Processes of the Human Mind: With a View to the Determination of the Fundamental Principles of Religions, Moral, and Political Science. A. Brown & Co., 1858.

Cronin, Anne M. Advertising and consumer citizenship: gender, images, and rights. Routledge, 2000.

Zizek, Slavoj. The plague of fantasies. Verso, 1997.

Zizek, Slavoj. The ticklish subject: the absent centre of political ontology. Verso, 2000.

Welton, Donn. Body and flesh: a philosophical reader. Wiley-Blackwell, 1998

Roof, Judith. Reproductions of reproduction: imaging symbolic change. Routledge, 1996.

Lucamante, Stefania. Italian pulp fiction: the new narrative of the Giovani Cannibali writers. Fairleigh Dickinson University Press, 2001.

Foerstel, Herbert N. Free expression and censorship in America: an encyclopedia. Greenwood Publishing Group, 1997.

Williams, Linda. Porn studies. Duke University Press, 2004.

McGowan, Todd. The real gaze: film theory after Lacan. SUNY Press, 2007.

Grant, Barry Keith. Film genre reader III. University of Texas Press, 2003.

Williams, Linda. Hard Core: Power, Pleasure and the 'Frenzy of the Visible'. Pandora Press, 1990.

Jermyn, Deborah; Redmond, Sean. The cinema of Kathryn Bigelow: Hollywood transgressor. Wallflower Press, 2003.

Jagodzinski, Jan. Youth fantasies: the perverse landscape of the media. Palgrave Macmillan, 2004.

Butler, Judith. Bodies that matter: on the discursive limits of 'sex'. Routledge, 1993.

Campbell, Jan. Film and cinema spectatorship: melodrama and mimesis. Polity, 2005.

Boer, Roland. Knockin' on heaven's door: the Bible and popular culture. Routledge, 1999.

Baudrillard, Jean; Zurbrugg, Nicholas. Jean Baudrillard: art and artefact. SAGE, 1997.

Pettman, Dominic. Love and other technologies: retrofitting eros for the information age. Stanford University Press, 2006.

Zizek, Slavoj. The indivisible remainder: an essay on Schelling and related matters. Verso, 1996.

Zizek, Slavoj. Mapping ideology. Verso, 1994.

Websites:

'shu lea Cheang' cityofwomen.org. 2001. 28 April, 2009. <http://www.cityofwomen.org/archive/2001/iku.html>

'Sex Machine Cams'Alan Stein. 2008. Butter Butter Productions Inc. 3 Jan 2009. < http://sexmachinecams.com>

'Fucking Machines', Peter Acworth. 1997-2009. Kink.com. 3 Jan 2009. < http:// fuckingmachines.com>

'Ki nk.com' Peter Acworth. 1997-2009. Kink.com. 3 Jan 2009. < http://kink.com>

'TriCaster Portable Live Production' 2009. NewTek. 3 Jan 2009. <http://www .newtek.com/tricaster>

Films:

I.K.U. Dir. Shu Lea Cheang. Uplink Co., 2001.

Thx 1138. Dir. George Lucas. American Zoetrope, 1971.

Sleeper. Dir. Woody Allen. United Artists, 1973.

Artificial Intelligence: A.I. Dir. Stephen Spielberg. Warner Bros. Pictures, 2001.

Flesh Gordon 2: Flesh Gordon Meets the Cosmic Cheerleaders. Dir. Howard Ziehm. New Horizons, 1989.

Barbarella. Dir. Roger Vadim. Paramount Pictures, 1968.

Demolition Man. Dir. Marco Brambilla. Warner Brothers, 1993.

Blade Runner. Dir. Ridley Scoot. Warner Bros.,1982.

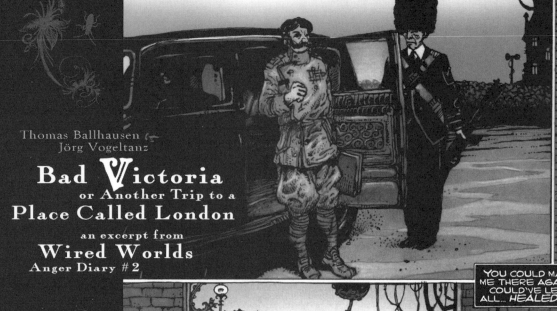

Thomas Ballhausen
Jörg Vogeltanz

Bad Victoria
or Another Trip to a
Place Called London
an excerpt from
Wired Worlds
Anger Diary # 2

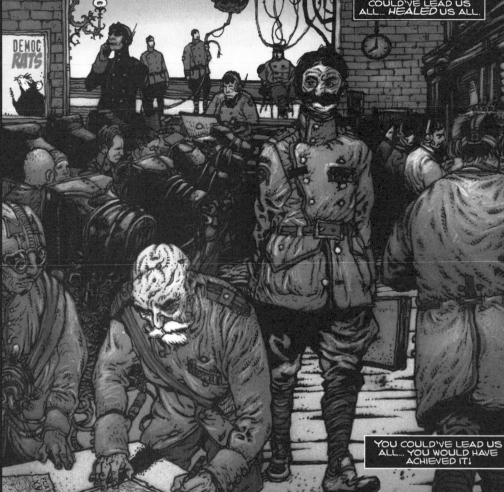

preQuel
www.prequel.at
anger.wetpaint.com

Tina Lorenz

A FUTURE IN PORN

Representations of sexuality are as old as humanity; in order to reinvent itself, this time-old genre has to explore new themes, go with the zeitgeist and have a bit of self-reflexive think about its own existence. Of course, movie porn did not start out as being expressly futuristic. In fact, stag film porn did not think about science fiction or visions of future societies at all. The stag film, privately shown in clubs to all-male, all-white, elite audiences was primarily busy showing voyeuristic scenes that were full of bawdy brothel humor or exoticism in its very short runtime. So instead of Ed-Woodian saucers on a string, spectators got peeps through keyholes, bathing beauties, or splendid harems.

A Free Ride, ca. 1915

A solid example of this is the ca. 1915 American Stag *A Free Ride*, which has been extensively analyzed elsewhere[1]. Suffice to say, it is firmly rooted in the contemporary here and now, with urination as the key to arousal, the car as status symbol and icon of sexual aggressiveness and virility and the representation of the women in dress and behavior as loose. Those features are characteristic of stags, in that they used mainly prostitutes for models, were shown only in male-dominated locations (such as brothels or exclusive clubs) and were made cheaply and secretly, demonstrated by the length of the reels used and the fact that intra-diegetic sound was introduced in the stag decades after it surfaced in Hollywood narrative cinema.

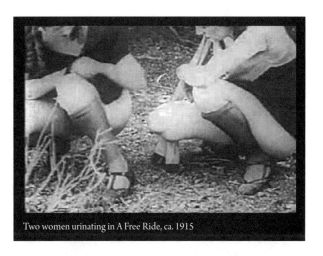
Two women urinating in A Free Ride, ca. 1915

Nude on the Moon, 1961

The first interest science-fiction roused in mainstream society was in the early 1960s, when Kennedy announced the race for the moon. Suddenly, space was the final frontier and sparked a multitude of productions in the realm of popular culture. Especially pulp magazines and (s)exploitation films in the 60s imaginatively knitted together narratives about space and sensationalism. One of sexploitation's pioneers was a woman – Doris Wishman. Until today she remains the most prolific female directors of all time with more than 40 films to her credit. Her experimentation with showing nude bodies in narrative settings for a young adult audience led to the creation of the nudie cutie, which features nudity, but no genitals, let alone hardcore action. Her most popular nudie cutie is probably 'Nude on the Moon', a 1961 flick about

Improvised astronaut gear in "Nude on the Moon", 1961

[1] Linda Williams in her seminal work Hard Core or Constance Penley in her essay 'Crackers and Whackers: The White Trashing of Porn' have both written extensively about this staple of American stag film history.

astronauts landing on the moon, only to find people living there in various stages of nakedness and surroundings that are oddly reminiscent of a nudist colony in Palm Springs (where the movie was shot).[2]

The space age together with the pioneer frontier spirit that went with it lent itself perfectly to a genre that was pushing its own frontiers at the time. Sexual imagery and mentions of indecency were, of course, extremely censored and frowned upon by the Hays Code that crippled Hollywood cinema from 1930 until 1968. Nonetheless, the sexploiters had found a loophole in the Code: education and documentation. If Wishman just documented life in a nudist colony, no one could complain. More often than not, spectators were pulled into the cinema by sensationalist posters promising them lewd images and were then bored by Wishman's very tame nudist flicks.

Flesh Gordon, 1974

After the end of the Hayes Code and the establishment of the film rating system, pornography experienced an atmosphere of departure that manifested itself in most of the films of that Golden Age. Porn was, even if only for a short time, chic. Average people came to see what all the buzz was about and as Gerard Damiano, director of 'Deep Throat' expressed it, porn had highest hopes of merging with Hollywood cinema at some point in the future[3]. Knowing this it is not surprising that porn took a turn towards Hollywood storytelling techniques: it actually centered on plot and its *mise-en-scéne* was modeled after narrative mainstream cinema. The ultimately unfulfilled wish to have a narrative cinema that does not halt

Phallic space ship in "Flesh Gordon", 1974

before the bedroom, but shows representations of human sexuality as well, led to the citation and copying of contemporary popular culture, so as to be more accessible and recognizable to the average movie goer. Thus, the subgenre of the spoof was created, which is now one of the staples of feature-length pornographic films. Many Golden Age porn films were lovingly handcrafted with wit and a good appetite for camp; finally, the Ed-Woodian visual effects took hold in porn and made such gems like 'Flesh Gordon' and his golden phallic space vessel possible.

Sculpted to resemble the comic super hero of the same sounding name, Flesh, his female mate Dale, and scientist Dr. Jerkoff steer their phallic space ship to the planet Porno, where the evil Emperor Wang has invented a sex ray that throws everyone in its proximity into frantic expressions of lust. With the ray, Wang aims at controlling and consequently destroying the humans through their sexdrive. Everything in this movie is over the top: the setting, plot, dialogue, character developments, even the sex scenes are prime examples of pure camp in the Susan Sontag (non-)definition: 'camp is not making fun of it; it is making fun out of it'[4]. For example in the scene where they just landed on Porno, knowing nothing about this strange planet, Jerkoff steps out of the spaceship, takes a couple of deep breath before turning to the others: 'Oh good, there's oxygen on this planet!'

[2] There is a brilliant documentary about (s)exploitation filmmaking in the 60s and 70s containing interviews with Wishman entitled 'Schlock: The Secret History of American Movies'.

[3] Damiano gave an interview for a documentary titled 'Inside Deep Throat'. This documentary film remains ambiguously received for various reasons, but I think that Damiano and many other pornographers of the 70s really believed in the validity and mainstream capability of what they were doing.

[4] Sontag, Susan. 'Notes on Camp'. Against Interpretation and Other Essays. Revised Ed. New York: Picador, 2001.

Cafe Flesh, 1982

Near the end of the cold war, when the world had been permeated with the scare of an atomic catastrophe and the AIDS epidemic was not a silent deadly illness killing hundreds without apparent cause but fear of STDs were rampant, disillusioned visions of dystopia entered the harmonic world of porn. The films of the 80s asked questions about the future of sexuality in the face of deadly diseases, and explored fearful visions of a society destroyed irreparably by those factors.

In *Cafe Flesh*, 99% of the people surviving an atomic world war get violently ill when they try to engage in sexual contact.

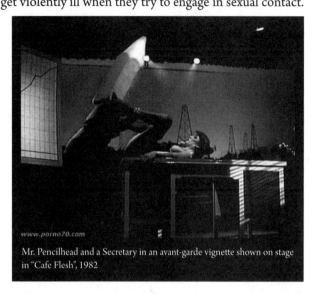

Only the remaining 1% are still living with their sexuality intact. These so-called 'sex positives' are compelled to perform onstage for the unhappy masses, whose only remaining pleasure it is to imagine what sex felt like. One female sex positive tries to remain hidden in order to stay with her sex negative boyfriend. Her emerging libido thwarts her plans and exposes her to other sex positives.

The ambitious and artsy *mise-en-scéne* of the film with its short vignette plays mimicking and slightly parodying the esthetics of the avant-garde embedded within the general narrative make *Cafe Flesh* a cult classic on the verge from the carefree Golden Age of Porn to the highly diversified Home Entertainment era. Its visions of dystopian art and exploring the theme of uninhibited expression of sexuality in a society where sexuality has acquired connotations of fear as the means to spreading STDs are still valid today.

Mr. Pencilhead and a Secretary in an avant-garde vignette shown on stage in "Cafe Flesh", 1982

Abducted by the Daleks/Daloids, 2005.

The urban myth that pornography helped VHS win the format war over Betamax is still perpetuated today; although there were many factors that caused VHS to win over Betamax (cheaper, longer tapes with which users could tape their favorite TV shows on one cassette and the successful lobbying of the major studios to put out old films on VHS rather

than Betamax cassettes, to name just two)[5]. Porn did figure in the equation insofar as it realized the potential of the new distribution system right away. The public outing of entering a sleazy grindhouse movie theatre and watching the movie in a social setting would be substituted with a considerably shorter exposure to potential awkward moments as the video was rented and returned to the video store. The actual reception of the movie went on in the privacy of the spectator's home, resulting in the emergence of so-called 'stroke flicks'

Bewildered girl and strangely impotent Daleks in "Abducted by the Daleks/Daloids", 2005

[5] An exhaustive discussion of this urban myth and its grain of truth can be found in Joshua Mark Greenberg's From Betamax to Blockbuster. Ithaca: Cornell University Press, 2004.

– films that center on sexual contact not plot and where most of the dialogue (if there is any) is geared to further the explicitness of the hardcore scene.

Another new aspect of the advent of VHS was the new control the user had. Fast-forwarding to the desired scene, replaying a favorite scene or even halting the film altogether was a revolutionary viewing technique that made new modes of production necessary: narrative was diminished in favor of longer sex scenes which were positioned at closer intervals, making the scenes more suitable for remote-controlled viewing. Subject matter from popular culture was also exploited systematically, often with dire results.

Thus films like the totally pointless *Abducted by the Daleks* were born. The 2005 film that had to be renamed *Abducted by the Daloids* (after the producers lost a lawsuit) features a handful of the vicious cyborgian mutants called Daleks that have seemingly arrived right out of classic sci-fi TV Show *Doctor Who*. The Daleks/Daloids discover a handful of blonde women in the woods who are in various stages of nakedness[6] and beam them onboard their spaceship. They start examining the girls and interrogating them with the help of a dominatrix. The aliens do have protruding instruments and such, but they constrict themselves to prodding and sticking the Dalek equivalent of a suction cup onto the girls' faces – yet no Dalek-on-girl action ensues. Eventually, the Daleks/Daloids release the girls. This renders the film unusual for the VHS age, as the focus is not on the sexual act itself, but in the difference between the (naked) female bodies and the machine-encased aliens. The erotic tension that is generated out of this opposition foreshadows developments in technology-aided expressions of sexuality (such as teledildonics and virtual sex) that predominates discussions about sexuality in the Internet age.

I.K.U., 2000

Intimacy with technology and technology-aided sexuality are also powerful themes of postmodern pornography. As the Internet slowly leads to declining DVD sales and new business models, a new type of pornography comes into being, together with new tropes and new modes of showing bodies in sexual ecstasy. One of tropes developed primarily in postmodern pornography is the sexual cyborg. As an afterthought of Haraway's 'Cyborg Manifesto' from 1991[7], the sexual

The cyborg as transhuman prototype in I.K.U., 2000

cyborg represents a closeness to humanity the humans themselves have long lost: in *I.K.U.*, Shu Lea Cheang's independent porn film from 2000, the cyborgs are actively pursuing the once thought to be essentially human quality of libidinous expression. They are collecting orgasmic experiences on their hard discs to be processed into a drug called I.K.U. This drug, once taken, leads the users to remember their most memorable orgasm as vividly as if they would go through the experience again – thus eliminating the need for real and possibly unsatisfactory sex. As the cyborg changes her appearance to accommodate the tastes of her respective partner, she is portrayed as a (post-)modern geisha, but with a twist: in a society where human sexuality (and thus, inevitably reproduction) is at an end, the sexual cyborg becomes the prototype of transhumanity and is only an orgasm away from a robot uprising.

---

[6] They got lost. Why one has to drop all clothes instantly when lost in the wilderness escapes my comprehension. But this film is not about comprehension. And the weird guy in the end that ties one of the girls to a tree in order to smear her body with lipstick? That's one weird endnote to the film.

[7] In the 'Cyborg Manifesto', Haraway argues that all humans could become or are in fact already cyborgs, and that cyborg sexuality in its singularity would actually refute the binary model of human sexuality, thus leading to a society in which sexuality as such is free.

The Future Of Visual Pornography

In the paragraphs above I have shown fair examples of pornography's interest in futuristic themes, visual experimentation with positions of the avant-garde and its own take on utopian or dystopian visions. What did not surface in this discussion was the question about the future of (visual) pornography as a part of popular culture. So what about porn's own future? As the meta-fictionality and self-referentiality that are staples of postmodern literature convention will increase and make its way into mainstream porn eventually, we may know more about what porn thinks about itself. For now, all is up to speculation.

Here are my favorite theses – they are of course at least partly powered by *Wunschdenken* – wishful thinking. Let's see if any of these will hold up in, say, 20 years:

The male gaze will diminish dramatically. As new audiences are being courted by various pornographic genres, the heteronormative, voyeuristic male gaze upon a female object of desire will be replaced with a diverse look upon people acting out their sexual desires before a camera. Heterosexual white men may still be the most avid consumers of pornography today – but who is to say if this will not change if the pornography offered is more inclusive of people not identifying with the gendered status quo? Also, feminist porn (the kind of pornography that establishes a necessarily choreographed sexuality in a diegesis free of heteronormative power games) will hopefully become such a strong force in pornography that it echoes back to other subgenres.

Visual effects and other techniques borrowed from Hollywood narrative cinema will be employed more. As modes of narration and new developments in visual technology have a tradition of arriving in the porn world a tad late, I would argue that we will see an increase of visual effects (either as instruments for visualization of the diegesis or as body doubles), especially in indie and narrative porn, as those sub-genres are apt to allow a significant amount of screen time to story development. The need for actual sets will therefore decrease.

Genres such as Altporn challenge traditional number-based pornography even today. This trend will probably continue and oppose the growing masses of uninspired fuckfests, thus diversifying the market even further.

We are currently standing at the crossroads of a change of paradigm. DVD sales are dwindling as new Internet formats are springing up. As of 2009, there is no clear overall strategy where all this is going to go, so all guesses are valid. I say, production on demand – for a real and diverse audience instead of an idealized white young male heterosexual spectator – could become a large factor. This direct interaction between producers and spectators and the possibilities of feedback call for a whole new take on pornography. What will happen if formerly passive spectators have the opportunity to give direct feedback and help shape pornography to their liking? This feeling of ownership and participation that permeates all of digital culture could dramatically change our perception of this staple of popular culture that is pornography.

Whatever is going to happen next, pornography hasn't spoken its last words about its own future, yet. As a genre with a rich and sometimes difficult history between violent persecution and uneasy acquiescence it will continue to be visible, controversial and transgressive – and with those functions continue to tell us more about ourselves than we probably wanted to know in the first place.

Cory Doctorow

MAKERS
EXCERPT FROM THE NOVEL MAKERS

This is the porniest sex-scene I've ever written, an excerpt from my October, 2009 novel MAKERS (published by Tor in the US and HarperCollins in the UK). Perry Gibbons is an inventor-turned-activist who has invented a kind of ride that celebrates the 'New Work,' a movement that encouraged people to up stakes and relocate to tech co-ops in dead malls and invent crazy things that were profitable for six weeks before they were cloned and something new had to be invented. Perry's rides are assembled by 3D printers, and anyone can contribute to them, and vote on their contents by working a little thumbs-up/thumbs-down lever in their ride vehicles.

Hilda and Perry just met tonight, while Perry was on his inaugural tour of the networked rides that tapped into his. She's a young, idealistic activist, second-generation, and a damned fine inventor in her own right (she created a line of papercraft furniture that is sold to students on the cheap). She and Perry have been out drinking and talking all night and now they've returned to Perry's little coffin hotel in Madison, WI:

'Man, I was really looking forward to spending a couple nights in my own bed.'

'Is it much more comfortable than this one?' She thumped the narrow coffin-bed, which was surprisingly comfortable, adjustable, heated, and massaging.

He snorted. 'OK, I sleep on a futon on the floor back home, but it's the principle of the thing. I just miss home, I guess.'

'So go home for a couple days after this stop, or the next one. Charge up your batteries and do your laundry. But I have a feeling that home is going to be your suitcase pretty soon, Perry my dear.' Her voice was thick with sleep, her eyes heavy-lidded and bleary.

'You're probably right.' He yawned as he spoke. 'Hell, I know you're right. You're a real smarty.'

'And I'm too tired to go home,' she said, 'so I'm a smarty who's staying with you.'

He was suddenly wide awake, his heart thumping. 'Um, OK,' he said, trying to sound casual.

He turned back the sheets, then, standing facing into the cramped corner, took off his jeans and shoes and socks, climbing in between the sheets in his underwear and tee. There were undressing noises - exquisite ones - and then she slithered in behind him, snuggled up against him. With a jolt, he realized that her bare breasts were pressed to his back. Her arm came around him and rested on his stomach, which jumped like a spring uncoiling. He felt certain his erection was emitting a faint cherry-red glow. Her breath was on his neck.

He thought about casually rolling onto his back so that he could kiss her, but remembered her admonition that they would not be having sex. Her fingertips traced small circles on his stomach. Each time they grazed his navel, his stomach did a flip.

He was totally awake now, and when her lips very softly - so softly he barely felt it - brushed against the base of his skull, he let out a soft moan. Her lips returned, and then her teeth, worrying at the tendons at the back of his neck with increasing roughness, an exquisite pain-pleasure that was electric. He was panting, her hand was flat on his stomach now, gripping him. His erection strained toward it.

Her hips ground against him and she moved her mouth toward his ear, nipping at it, the tip of her tongue touching the whorls there. Her hand was on the move now, sliding over his ribs, her fingertips at his nipple, softly and then harder, giving it an abrupt hard pinch that had some fingernail in it, like a bite from little teeth. He yelped and she giggled in his ear, sending shivers up his spine.

He reached back behind him awkwardly and put his hand on her ass, discovering that she was bare there, too. It was wide and hard, foam rubber over steel, and he kneaded it, digging his fingers in. She groaned in his ear and tugged him onto his back.

As soon as his shoulders hit the narrow bed, she was on him, her elbows on his biceps, pinning him down, her breasts in his face, fragrant and soft. Her hot, bare crotch ground against his underwear. He bit at her tits, hard little bites that made her gasp. He found a stiff nipple and sucked it into his mouth, beating at it with his tongue. She pressed her crotch harder against his, hissed something that might have been *yesssss*.

She straightened up so that she was straddling him and looking imperiously down on him. Her braids swung before her. Her eyes were exultant. Her face was set in an expression of fierce concentration as she rocked on him.

He dug his fingers into her ass again, all the way around, so that they brushed against her labia, her asshole. He pulled at her, dragging her up his body, tugging her vagina toward his mouth. Once she saw what he was after, she knee-walked up the bed in three or four quick steps and then she was on his face. Her smell and her taste and her texture and temperature filled his senses, blotting out the room, blotting out introspection, blotting out everything except for the sweet urgency.

He sucked at her labia before slipping his tongue up her length, letting it tickle her ass, her opening, her clit. In response, she ground against him, planting her opening over his mouth and he tongue-fucked her in hard, fast strokes. She reached back and took hold of his cock, slipping her small, strong hand under the waistband of his boxers and curling it around his rigid shaft, pumping vigorously.

He moaned into her pussy and that set her shuddering. Now he had her clit sucked into his mouth and he was lapping at its engorged length with short strokes. Her thighs were clamped over his ears, but he could still make out her cries, timed with the shuddering of her thighs, the spasmodic grip on his cock.

Abruptly she rolled off of him and the world came back. They hadn't kissed yet. They hadn't said a word. She lay beside him, half on top of him, shuddering and making kittenish sounds. He kissed her softly, then more forcefully. She bit at his lips and his tongue, sucking it into her mouth and chewing at it while her fingernails raked his back.

Her breathing became more regular and she tugged at the waistband of his boxers. He got the message and yanked them off, his cock springing free and rocking slightly, twitching in time with his pulse. She smiled a cat-ate-the-canary grin and went to work kissing his neck, his chest - hard bites on his nipples that made him yelp and arch his back - his stomach, his hips, his pubes, his thighs. The teasing was excruciating and exquisite. Her juices dried on his face, the smell caught in his nose, refreshing his eros with every breath.

Her tongue lapped eagerly at his balls like a cat with a saucer of milk. Long, slow strokes, over his sack, over the skin between his balls and his thighs, over his perineum, tickling his ass as he'd tickled hers. She pulled back and spat out a pube and laughed and dove back in, sucking softly at his sack, then, in one swift motion, taking his cock to the hilt.

He shouted and then moaned and her head bobbed furiously along the length of his shaft, her hand squeezing his balls. It took only moments before he dug his hands hard into the mattress and groaned through clenched teeth and fired spasm after spasm down her throat, her nose in his pubes, his cock down her throat to the base. She refused to let him go, swirling her tongue over the head while he was still super-sensitive, making him grunt and twitch and buck involuntarily, all the while her hand caressing his balls, rubbing at his prostate over the spot between his balls and his ass.

Finally she worked her way back up his body licking her lips and kissing as she went.

'Hello,' she said as she buried her face in his throat.

'Wow,' he said.

'So if you're going to be able to live in the moment and have no regrets, this is a pretty good place to start. It'd be a hell of a shock if we saw each other twice in the next year - are we going to be able to be friends when we do? Will the fact that I fucked your brains out make things awkward?'

'That's why you jumped me?'

'No, not really. I was horny and you're hot. But that's a good post-facto reason.'

'I see. You know, you haven't actually fucked my brains out,' he said.

'Yet,' she said. She retrieved her backpack from beside the bed, dug around it in, and produced a strip of condoms. 'Yet.'

He licked his lips in anticipation, and a moment later she was unrolling the condom down his shaft with her talented mouth. He laughed and then took her by the waist and flipped her onto her back. She grabbed her ankles and pulled her legs wide and he dove between her, dragging the still-sensitive tip of his cock up and down the length of her vulva a couple times before sawing it in and out of her opening, sinking to the hilt.

He wanted to fuck her gently but she groaned urgent demands in his ear to pound her harder, making satisfied sounds each him his balls clapped against her ass.

She pushed him off her and turned over, raising her ass in the air, pulling her labia apart and looking over her shoulder at him. They fucked doggy-style until his legs trembled and his knees ached, and then she climbed on him and rocked back and forth, grinding her clit against his pubis, pushing him so deep inside her. He mauled her tits and felt the pressure build in his balls. He pulled her to him, thrust wildly, and she hissed dirty encouragement in his ear, begging him to fill her, ordering him to pound her harder. The stimulation in his brain and between his legs was too much to bear and he came, lifting them both off the bed with his spasms.

'Wow,' he said.

'Yum,' she said.

'Jesus, it's 8AM,' he said. 'I've got to meet with Luke in three hours.'

'So let's take a shower now, and set an alarm for half an hour before he's due,' she said. 'Got anything to eat.'

'That's what I like about you Hilda,' he said. 'Businesslike. Vigorous. Living life to the hilt.'

Her dimples were pretty and luminous in the hints of light emerging from under the blinds. 'Feed me,' she said, and nipped at his earlobe.

In the shoebox-sized fridge, he had a cow-shaped brick of Wisconsin cheddar that he'd been given when he stepped off the plane. They broke chunks off it and ate it in bed, then started in on the bag of soy crispies his hosts in San Francisco had given him. They showered slowly together, scrubbing one-another's backs, set an alarm, and sacked out for just a few hours before the alarm roused them.

They dressed like strangers, not embarrassed, just too groggy to take much notice of one another. Perry's muscles ached pleasantly, and there was another ache, dull and faint, even more pleasant, in his balls.

Once they were fully clothed, she grabbed him and gave him a long hug, and a warm kiss that started on his throat and moved to his mouth, with just a hint of tongue at the end.

'You're a good man, Perry Gibbons,' she said. 'Thanks for a lovely night. Remember what I told you, though: no regrets, no looking back. Be happy about this - don't mope, don't miss me. Go on to your next city and make new friends and have new conversations, and when we see each other again, be my friend without any awkwardness. All right?'

'I get it,' he said. He felt slightly irritated. 'Only one thing. We weren't going to sleep together.'

'You regret it?'

'Of course not,' he said. 'But it's going to make this injunction of yours hard to understand. I'm not good at anonymous one night stands.'

She raised one eyebrow at him. 'Earth to Perry: this wasn't anonymous, and it wasn't a one-night stand. It was an intimate, loving relationship that happened to be compressed into a single day.'

'Loving?'

'Sure. If I'd been with you for a month or two, I would have fallen in love. You're just my type. So I think of you as someone I love. That's why I want to make sure you understand what this all means.'

'You're a very interesting person,' he said.

'I'm smart,' she said, and cuddled him again. 'You're smart. So be smart about this and it'll be forever sweet.'

Susan Mernit / Viviane

AVOIDING THE EMILY GOULD EFFECT
SEX, BLOGGING, NARRATIVE AND TRANSPARENCY

'Of course, some people have always been more naturally inclined toward oversharing than others. Technology just enables us to overshare on a different scale.'
Emily Gould, 'Exposed', New York Times Magazine, August 2008

Back in 2007, when ambitious blogger Emily Gould turned her dating life into a blog post-by-blog post reality show in search of Internet micro-fame (and notoriety in the pages of Gawker), we didn't yet get that the web - with its endless, insatiable appetite for sensationalism, porn and personal stories - was going to be crack for famewhores. Whether you were a small town exhibitionist or a world-class fame-chaser, digital media - and the blogosphere in particular - offered an irrestible platform to share, overshare and show off.

For Emily Gould, age 24 and about to become co-editor of Gawker, the snark/media blog (and this moment's Web 2.0 Tatler), the opportunity to pander to the crowd by ratting out her love life was too hard not too take. 'I am going to try not to write about you,' she breathed to the boy in her bedroom one night, promptly breaking that rule (as, in retaliation, so did he.)

Of course, none of this would have meant much of anything if it hadn't, in August 2008, led to a provocatively photographed cover story in the New York Times Magazine in which Gould simultaneously dishes all the dirt on herself and her naughty ways and whines about how she Ruined Her Life.

> 'I didn't want to exist,' Gould told the readers of the New York Times. 'I had made my existence so public in such a strange way, and I wanted to take it all back, but in order to do that I'd have to destroy the entire Internet. If only I could! Google, YouTube, Gawker, Facebook, WordPress, all gone. I squeezed my eyes shut and prayed for an electromagnetic storm that would cancel out every mistake I'd ever made. '

Poor Emily Gould! And lucky New York Times Magazine! And lucky us, who get the vicarious thrill of both Gould's salacious account and of her prettily teared regret.

Welcome to the world of TMMI - too much micro-information sharing - and an age when women twittering from the delivery room an hour after birthing their child (and twittering from the OB-GYN's table during an exam) are not regarded as uncommon.

So, in the world of TMMI - where blog posts, Twitter streams, YouTube videos and Flickr photos - among other things - often give you the inside blow by blow on the personal details of people you have never met (and who are * famous * to someone, even if it's only themselves) what are the rules and best practices that can keep you from being viewed as a famewhore and micro-slut, jiggling your juicy bits on the net in a vain bid for respect (a different quality than attention)?

How do you avoid the overexposure - and in some cases, ridicule - that bloggers including Julia Allison, Emily Gould, Xeni Jardin, and Zoe Margolis have endured?

Was making the August 2008 cover of Wired as the Paris Hilton of the Net what Julia Allison imagined would happen when she (first) came out to Silicon Valley and a TechCrunch party in summer 2007? Did she imagine that her videos on Vimeo with then-boyfriend Jakob Lodwick would snag ridicule rather than respect? (And would it have been different if a) she had a longer online history as a real digerati (she's was a magazine writer) and b) was a guy?)

For Xeni Jardin, a co-editor of the popular web zine BoingBoing, notoriety struck when she removed all the posts by a former friend (and lover) San Francisco sex blogger and educator Violet Blue from the popular site. The subsequent outpouring of reader protest and media coverage exploded as netizens protested Jardin's 'disappearing' Blue's posts and Jardin refused to talk about her reasons. Privacy=FAIL.

For Zoe Margolis, aka Abby Lee, there was probably no way to avoid the mess that happened. Margolis first blogged about sex as the Girl with a One Track Mind in 2003 in the United Kingdom, where she lived. In 2006, she published a book based on her blog and got outed by the Times of London. With the resulting media circus, life as she knew it crumpled. Zoe changed her job, moved to the States, and continued to blog, but has said that her dating life has been forever colored - badly - by her fame.

> '…I lost all trust in dating, and men in general, removing myself from the dating arena entirely,' she wrote in 2008. 'The months after my 'outing' in the press were spent mostly on my own - ironic for such a previously 'active' sex diarist.'

So, readers, here's what we want you to think about:

Are these women who blogged about their sex lives punished for being 'sluts?'
Is it that a guy writing about his coke habit okay, but a woman writing about her sexuality isn't?
Or are we all sex-phobic?
How could this story be different?

There's no question but that there are other sex bloggers - and feature writers - who have done a great job skirting the fine lines of sexy micro-celebrity. Former sex worker - and Valleywag alum Melissa Gira Grant, former Village Voice sex columnist and lusty lady Rachel Kramer Bussel, New York Magazine writer Rex Sorgatz, and Gawker writer Nick Douglas are among the blogging - and twittering - micro-celebrities who seem to have managed to reveal much and yet not end up completely trashed. Or, to put it another way, they each managed to finesse their over-sharing so it worked to their advantage, getting them Twitter followers, media attention and sometimes, writing gigs, without apparently disrupting their personal lives in the ways that Gould and Lee experienced.

What made Rachel Kramer Bussel, Nick Douglas and the others succeed, in our opinion, was a shared ability to draw a boundary and maintain it. As Nick Douglas said, '… A professional writer can't overshare, they can only waste good material or put something out there before it's polished enough' meaning having control of your persona, and what you give away, is essential.

So, if you're going to blog about your personal life, your sex partners, your erotic fantasies, what is the agreement you have to make with yourself for your narrative to succeed?

What are the rules that will help you keep your work defined? For some bloggers it's as simple as:
Don't post in anger.
Don't blog drunk.

But there is also the question of voice.
Is fisking your art form?

Do you snark, bigtime?

And how much do you actually want to share?

Reading Emily Gould's accounts in her New York Times Magazine cover story of trading online barbs with former boyfriend Joshua David Stein (who himself outed their relationship in a much-discussed piece in Page Six Magazine (http://www.nypost.com/seven/05232008/entertainment/the_dangers_of_blogger_love_112227.htm), we are reminded of nothing so much as articles the New York Times Magazine had previously published that were equally salacious - notably Joyce Maynard's revelations that she wrote to J.D. Salinger, went to his farmhouse and then spent some time as his much younger sweetie (he was 53, she was 18) and Elizabeth Wurtzel's revelations of her life as a suicidally depressed Harvard undergraduate and girl about town.

Working both sides of the issues, anyone?

As bloggers who have long been interested in issues of privacy and identity, and as writers about sexuality and relationships, there's no way we can't feel respect and compassion for Gould's struggles.

And yet, at the same time, as people with life experience and history, we also can't help feeling that running this story amounts to some editor at the Times taking a calculated, sensationalistic assessment of Gould's looks and persona and deciding to put that kettle to boil as a way to make the paper of record seem hipper, more relevant, and a place where buzz happens (yellow journalism be damned).

Or, to put it another way, as much as we are uncomfortable with the ways these writers describe how each of them acted in the pursuit of fame, fortune, audience and approval, we are more uncomfortable with the Times' consistent publication of stories - like Emily Gould's - that pretend to discuss high brow issues, but are really just sensationalist reads that give Sunday magazine readers a chance to live vicariously.

There's a ton of commentary in the blogosphere on Gould's piece, all worth a read. At Gawker, commentators Cassandra and The Dismal Science, elegantly debated whether Emily Gould's self-obsession reflects her own personality or her entire generation's narcissism:

> 'And none of this is as bad as the fact that her writing, initially, seemed genuine, raw, and destined for something larger. Instead, she's already hooker her perpetual media motion device meager dark energy of her 'persona' and doesn't seem to be concerned that, absent intervention, the height of her journalism career might be the time she dueled her ex-boyfriend's article with her own.'
> (http://gawker.com/392697/we-are-all-emilys)

In the Huffington Post, Rachael Sklar, writing an essay called Emily Gould: New Gloss on an Old Story, says:

> 'It's the NYT's call, to be sure, but I can't help thinking they got snowed; it's the third magazine piece on Gawker, and the second on the star-cross'd Gould-Stein hookup.'
> (http://www.huffingtonpost.com/2008/05/23/emily-gould-new-gloss-on_n_103241.html)

For Israeli feminist blogger Miriam Schwab, the story is all about privacy and persona:

'Her piece is fascinating and disturbing, and raises a lot of questions about the boundaries we set up and break down between our real-life identities, and those of our online personas.'
(http://illuminea.com/blogging/emily-gould-gawker-privacy-publicity-web/)

Erica Perez notes the comments and the commotion Gould's story caused:

'The comments were the most interesting part. By Wednesday evening when I read the story, it had more than 800 comments, 90% of which were scathing, criticizing Gould for being a narcissist, an idiot, a bad writer and a horrible person, and taking the NYT to task for running the story so prominently when, for example, people are suffering in China and Iraq.'
(http://blogs.jsonline.com/fishoutofwater/archive/2008/05/26/nyt-story-on-blogging-overexposure-and-well-the-world-today.aspx)

While Megan McArdle writes perceptively in the Atlantic:

'Gawker both expanded her horizons and terribly limited them; from the perch of her overflowing inbox, she could see everything in the world (or at least Manhattan). Yet quickly enough she became the only thing she cared about within it. The entire city of New York mattered only insofar as it was a reflection of Emily.'
(http://meganmcardle.theatlantic.com/archives/2008/05/the_saga_of_emily_gould_1.php)

For us, Gould's tale is that double-edge sword of caution and success, fame whoring and putting it out there. The deconstruction of her over-sharing is both the admiration for her book contract and the loss of reality her public pandering engendered; the lessons for all of us, in this age of over-sharing, is that putting it out there is a pleasure - but only if you maintain control of your own persona.

Mela Mikes

WHAT'S LOVE GOT TO DO WITH IT?

What is love but a second hand emotion? Remember when talking about love was talking about desire and talking about desire was talking about feminism and talking about feminism was talking about differences? During my most recent readings of political theories it felt like those were the good old days. What happened? Honestly, I am not a hundred percent sure about that.

It will not be possible to talk about all we should consider when talking about political agency, gender, sex and politics. Additionally there is something – and you might have come across this yourself – everytime Zizek readings are involved it gets blurry and hard to grip or even harder to 'explain', but maybe that is giving him too much credit. Maybe this happens everytime language and meaning is talked about. Maybe that's it. Right, you'll say - but we're talking about love, aren't we? Love is often mistaken as 'only' a feeling, chemistry or even nature. It is not exactly that. Love is a cultural concept, a convention, something that is of course bound to the logic of language.

Language - both Zizek and Badiou are using post-structural concepts of language. Theirs is derivated from Lacanian psychoanalysis. It would really be overkill to talk about Lacan's ideas in detail. But it would be even more confusing to not discuss a few Lacanian ideas since they are the very foundation of both Zizek's and Badiou's views. When Lacan expresses that there is no connection to the 'Real' itself then Zizek would say there are moments when the real breaks through and a connection is possible.

When Lacan says that the 'economics' of love are based on an exception and that there is no way to really get the Other as the complete other without thinking and desiring yourself in it then Badiou is forcing to make this exception a new universalism.

So far, so good. Love is the message and the message is love.
These theories are situated in post-empire theory and post 9/11. To me it is very interesting to see how political agency is being shifted away from the complicated yet open settings of the theories of the 1990s to somehow closed and desperate ones in recent publications. What I mean is, that if we take a closer look at the writings of Zizek and Badiou in particular we can see how a pledge for 'universalism' can go wrong although there have been similar pledges in the early nineties by feminists like Donna Haraway that are more successful.

Arguing for oneness, for united stands and equal voices, a unified goal is a very strong argument, so it seems. It was not a small part in the ongoing discussions in the late 1960s up to the early 1990s to find ways to involve the silent masses of the oppressed and unpriveledged. It was this oppression that should unify and create the sense of solidarity and agency to overcome repression. Materialist analysis puts its emphasis on exactly that point. Of course a 'common sense' of morality should tell us that it is just natural to put sympathy and solidarity to those who are not fortunate. Ironically the summer of '68 that took place in spring right here in San Francisco was flooded with future lawyers and professors that could go back to their trust funds and easily switch back to the elitist benefits of the establishment they tried to fight against.

Does this fact make their claims less valid? Yes and no, and that is another form of the problem a claim for 'universalisms' always bares. As the effect of the struggle doesn't mean the same to all involved and that effect is made invisible within the claim for 'universalism'. What is the problem or need that leads to pledging for universalism?

It is the apparent goal of salvation from capitalism, which could be translated into the more abstract goal of a paradise that is waiting for 'us'. This paradise is a promise that comes in many disguises - may it be nirvana or heaven, the messiah or peace on earth, no pollution or simply nature undisturbed by humankind.

One of the many questions coming with the craving for paradise is who's in and who is out? Which is all but an ethical discussion. The paradise or utopia might come as something worth to put yourself at risk for. It has to be desirable and it has to be within reach. Here 'within reach' means that the way to get there has to be sketched out, more or less.

Funny enough, the sheer concept of paradise alone doesn't do anything. You can always say that's not for me, but maybe it is for someone else. That is where 'universalism' comes into play once more and with more force. Who will be in and who'll be out? 'When I am king you'll be the first against the wall ....' is one way to solve the problem. Another one is to say that there is no need for a binary structure to this. It would be a pledge for situated objectivity that doesn't need to be re-assured with 'one'.

It is not very surprising that Alain Badiou is using St. Paul as the figure which is embodying his idea of a 'new universalism'. He states in his book on Paul that he was the first one to communicate a singularity, which would be the resurrection of Jesus. For Badiou the story of Saul who becomes St. Paul is the story of how to overcome an established order and establish a new one. Not only that but also how to involve 'masses' to this new order by offering ways to participate in the 'truth' event.

It might not sound very spectacular to you but if you take another minute and think about it, there is someone offering you something that is in itself 'true' that means it is free of any doubt and any responsibility. In addition it is good and will guide you to paradise. I'll have to say that of course Badiou is not making it that simple and he wouldn't want to have it seem simple, since he puts a lot of effort into his logic of the 'supernumerary', which means that the one is always a partial one that is split in itself. So what the universal Badiou is trying to create is folding from and between this 'supernumerary' subject.

Indeed this comes rather close to what Donna Haraway describes as situated knowledge in her claim towards a new objectivity. Yet, she wouldn't want to have this objectivity ultimately translated into one truth or paradise, nor one powerful agency to achieve it. And that is where Love comes in. Badiou puts an emphasis on love and grace. This love is the love of the mother to her child. Since the father is the law that we should overcome. Hence this is why Jesus has to provide the truth event. A truth event is not only providing truth it is also true in itself and has no contradiction. Christian love is not the not hate of god, it is love and love alone. That means it cannot provide anything false or evil.

The ultimate agency is represented through a fatherless child unconditionally loved by its holy mother that is free of any original sin. Kierkegaard describes 'fear' as the root of the original sin. The concept of the original sin is bound to the concept of the re-occurring of the ever same as it is reborn with every new human. So Mary was without any fear of the future. This would be a rough capturing of what 'unconditional' means in this context. It is obviously the case that only those who subscribe to the new universalism will be loved unconditionally. The rest might as well just be ignored, if they're lucky. This notion of unconditional love is what should enable to overcome 'the law'.

This 'abstract' Law is what Badiou and Zizek identify as the evil to overcome. Throughout my readings it felt like law and capitalism were used interchangeably.

This points to a slightly anti-Semitic tone especially within Zizek's writings, as the law is not only abstract but equalized with Judaism. Another thing that is attached to the law is history. Alain Badiou not only wants to say that political agency needs a 'new' universalism, no, it also should provide the possibility to transcend history, not only the personal one but also the collective history.

The new order would be free of the old law, and would no longer be attached to the old language and meaning. Doesn't that sound just great? All that achieved just by love for the real event, the singularity, a break.

Yet who would we be without history, without guilt, without the burden of meaning? I have to confess that the idea of such a political agency scares me more than it liberates me. It was the love for transcendence that generated the worst in humankind throughout history. Purity as a consequence is attached to this agency, so either you believe or you don't believe. It is paranoid in itself and in its future.

It seems like as if Zizek is struggling with this part of Badiou's theory, since he is putting emphasis on 'the death of Jesus on the cross'. This allows him to introduce 'doubt' into his version of christian love. Whereas doubt has not much place in Badiou's theory, it becomes more and more important in Zizek's. Zizek, following Mouffe's and Laclau's concept of political agency, is very much against any equations of institutions and actual practices. He moved from a harsh critique of Badiou to finally introducing large parts of Badiou's ideas into his own. Zizek combines Leninism and Christianity with the ultimate political agency.

This one should enable a transparent self. Something that is impossible for Freud and Lacan. It literally would be a self that can read the unconscious without any problem, or more precisely, it would locate the unconscious within the self. It is a self that shares space with the real, and hence becomes nature minus any need for language or history. For a long time Zizek doesn't dare to go this far in his writings, although in his recent publications he suggests that this transparent self is what the dying Jesus on the cross resembles. Dying in love for our sins without any grim against Judas, he doubts for a moment asking 'oh my god, why did you leave me'. This is indeed spectacular.

When he puts the possibility of connecting with the real (the world as it is without the need of translation) he creates a psychotic subject, one that is full of visions and free of reality and fulfillment of the law. Both, Zizek and Badiou go with the Paulian paradigm of 'love is the fulfillment of the law'. And in this very statement it becomes clear why they put their emphasis on love as the foundation of their new universalism.

As a feminist and as someone who went through different stages of resistance against hegemonial power it makes me very suspicious to read such suggestions of political practice. Honestly I wouldn't even put a thought into reading that stuff, if those two weren't good examples on the backlash in political theory and activism. At least in the discussions in Europe and especially in the German speaking parts of it, Zizek and Badiou are the heroes of middle aged white activists.

And with these elaborations you may understand why. After the years of Simulakra, Rhizom and post-feminism it must be wonderful to subscribe to something that provides a new common ground. But sad to say, it is not only a claim for a common language, like the feminists were struggling for. No, it also is a claim for a common truth and as a consequence a common ethical codex. You might say - but there is the fractual self Badiou provides, plus the doubt that Zizek adds. But isn't all that nothing more than a likeness of a very conservative, even racist theory since those little concessions are becoming almost invisible within the suggestion of one paradise for all who believe? Like the marxist activists in the student movements Badiou and Zizek would say that racism and sexism are no struggles themselves as they'll be solved once the law is ineffective.

And that makes love not enough.

It might not be as catchy and comfortable to think political agency along an even more fractured, less assuring path. When in 1985 the Manifesto for Cyborgs was published, Donna Haraway was embarking upon a very unusual path. She was suggesting that we don't have to decide whether or not we're women by nature, since we're all cyborgs. She was clearly speaking out against the promise of a feminist paradise that will await us at the end of our struggles. She was one of the first to tell us that we are not united by the purported nature of being women. Others like Rosi Braidotti or Teresa de Lauretis have offered insights into building new narratives on ethical questions of what constitutes a subject.

Therefore I was more than surprised to see the paradise popping up again in the theories of middle aged male philosophers. While I can understand the desperation that comes with reading post structural theory, I find comfort in thinking that there is not one sole truth and not a right or a wrong thing to do. Isn't it comfort enough to know that trying alone is worth trying and that failure is not taking away any validity of the effort?

Like Susanne Lummerding highlights, terms like truth and validity are not preexistent, they are created within the process of negotiation, resistance and the fight for ideas. But not everything can be valid - you'll say. And yes, that is true: this kind of subjectivity is not the opposite of objectivity. It took me 10 years of reading to get to the point to actually understand that kind of logic. You may be faster of course, I was not.

An argument that needs the postulate of universalism to its claims as a precondition for validity is of course not able to reflect a structure that creates validity in the translations of its contents between the points/knots of the structure. That is why it is always a question of the method that has an effect on the ethics of an argument/theorem. But that is exactly what guys like Badiou and Zizek would deny in their claims of universalism. Since there is no concept of what the truth event means to those who don't believe, universalism is only unfolding for those who believe.

This neglecting of minorities is what makes their ideas invalid for me. A call for political agency that is meaningful for everyone, that will simultaneously lead to salvation, is not really something I'd like to subscribe to.

Maybe that is why reading Giorgio Agamben's 'the time that remains' was more meaningful to me. His precise reading of Badiou and the Paulanian letters especially (the ones to the Romans) is pointing to the problem of taking history away form the subjects and also points to the problem of recognising when said paradise is starting. This probably sounds like an easily answered question. It's like most people say they'd change everything if they could. My typical response is that if they had the chance they'd actually change nothing, since change is a parameter that evolves by shifting something, not in erasing it from the surface.

It is a shame that it's impossible to hand over all the interactions these theories have with other hegemonic or 'progressive' political theory. I hope that some of you sense the need to ask about preconditions and offers within the ideas you use to cope with society as it is.

As you can see with all the points I've raised these discussions are very extensive. I don't want to hurt anybody's religious feelings, but I really don't want to see religion in a place where it is describing so called 'progressive' politics. I don't want to have the option to either believe or not to believe. And I don't feel like 'love' or any other feeling should be the practice for a new 'universalism' since I do not see how such a new universalism would do any good.

I beg your pardon, but I never promised you a rose garden.

This talk is based on following readings:

Badiou, Alain: Paulus. Die Begründung des Universalismus. formerly sequenzia   Verlag, now diaphanes 2002.engl.'Saint Paul: The Foundation of Universalism. Standford: Standford University Press, 2003

Slavoj, Zizek: The Parallax View. The MIT Press, Cambridge 2006

Slavoj, Zizek: Das Reale des Christentums. Suhrkamp, Frankfurt 2006

Slavoj, Zizek: The puppet and the dwarf. The perverse core of christianity. The MIT Press,Cambridge 2003

Slavoj, Zizek: On Belief. Routledge, London-New York 2001

Slavoj, Zizek: Haraway, Donna: Ein Manifest für Cyborgs, Die Neuerfindung der Natur. Campus Verlag 1995. engl: Cyborg Manifesto in Simians, Cyborgs and Women: The Reinvention of Nature. RoutledgeNew York and London, 1989

Lacan, Jaques: Une lettre d'amour in The Seminar XX, Encore: On Feminine Sexuality, the Limits of Love and Knowledge ed. by Jacques-Alain Miller, transl. by Bruce Fink, W.W.Norton & Co., New York, 1998.

Gilles Deleuze: Die Logik des Sinns. Suhrkamp,Frankfurt a.M.1989, orig:'Logique du sens (1969). engl: trans: The Logic of Sense. Columbia University Press, 1990

Giorgio Agamben : Die Zeit die bleibt. Ein Kommentar zum Römerbrief. edition suhrkamp, 2006. engl: The Time That Remains: A Commentary On The Letter To The Romans.Stanford Univ Press,2005

Annalee Newitz

CAN A ROBOT CONSENT TO HAVE SEX WITH YOU?

It's a truism in adult science fiction that humans of the future will have sex with robots. But can a robot really consent to have sex when it's been programmed?

Under the law, the difference between an act of sex and an act of assault hinges on one idea: consent. If a person agrees to have sex with you, you're having sex. If they don't agree, or actively disagree, it's a crime. Obviously there are gray areas, and that's why rape trials exist - in the best cases, such trials are intended to determine whether consent was given.

But what about robots? Do you think the blondie bot in *Cherry 2000*[1] was really capable of giving consent to have sex with her human boyfriend? Or did her programming simply force her to always have sex, whether she wanted to or not? And what about the Romeo Droid in *Circuitry Man*[2], or the Sex Mecha in *A.I.*[3], who live entirely to sexually please women, even when those women are abusing them or putting them in danger?

Then there's the opposite problem, which Ekaterina Sedia tackles in her recent novel *Alchemy of Stone*[4]. Her main character is a robot whose creator built her without genitals. Even when she wants to have sex, her body makes it impossible for her to consent in a recognizable way (though she does manage to figure out a technical workaround).

Whether programmed to have sex, or designed to refuse it, the problem these fictional bots face is a lack of control over their own desires. You can't really be said to consent to sex if you're never given the option to choose between 'yes' and 'no.'

Cat Rambo[5] has written a short story where one of the characters is a female superhero whose mad scientist creators made her hyper-sexual. No matter what happens, she's always aroused, regardless of whether she wants to be or not. Her solution to this design feature is never to have sex with anyone. She doesn't like the idea of being trapped inside a sexual desire that a bunch of men designed into her without consent.

Researcher David Levy got a lot of media attention[6] for his recent book *Love and Sex with Robots*, where he argues that by 2050, people won't just be having sex with bots - they'll be falling in love with them, and even marrying them. He talks about the development of emotional and social robots, creatures programmed to perceive and imitate human emotions. Already, roboticists at MIT have created several models of bot that respond to facial expressions and tone of voice with so-called appropriate emotions: An angry voice makes the bot cower; a smile returns a smile.

But of course these emotional robots have been programmed with what somebody thinks is an appropriate response - sort of the way Rambo's superhero has been programmed to respond to everything with sexual arousal. If we accept that robots will achieve human-like intelligence, it seems likely that such bots will sense a difference between what their programming makes them do and what they actually want to do.

[1] http://io9.com/326667/before-human+cylon-love-there-was-a-dude-and-his-cherry-2000

[2] http://www.imdb.com/title/tt0099271

[3] http://www.imdb.com/title/tt0212720

[4] http://io9.com/5027493/a-living-doll-tries-to-survive-a-workers-revolution-in-the-alchemy-of-stone

[5] http://www.kittywumpus.net

[6] http://www.nytimes.com/2007/12/02/books/review/Henig-t.html?_r=2

So if a robot has been programmed to respond to human sexual arousal with more sexual arousal of its own, is he actually consenting? Or is he just going through the motions of pleasure and desire, wishing that he could control his own responses enough to choose whom he had sex with, and when?

Questions like these, raised in science fiction or speculative science writing like Levy's, are inevitably really questions about ourselves. As of yet, we have no bots who are sophisticated enough to experience intimate relationships with humans - by programming, or by choice. But as humans, we often exist in the gray areas of consent when it comes to sex. Our physical desires, our basic sexual programming, may conflict with what we actually want to do.

Certainly there are many situations where it is obvious that consent has not been given, or has been. But for all the situations in the middle, we are like the bots we imagine that one day we will fall in love with. We cannot untangle what we think we should do (our social programming) from what we want to do. Or we can't disengage our raging physical urges (more programming) long enough to ask, 'Wait, do I really want to have sex with this person? Or do I just want to have sex with anything, including furniture?' In Charles Stross' excellent novel *Children of Saturn*, the always-randy sexbot heroine knows the answer to this question, and responds by humping hotel rooms and spaceships.

So will you ever be able to have consensual sex with a robot? Maybe. Sometimes. Unless you aren't bothered by having sex with a slave or a brainwashed victim, having relationships with robots will probably be just as complicated as having them with humans.

This is the first in a series of columns called Fully Functional[7] *that I'll be writing about science fiction and sex. If there are any topics you want me to tackle, pipe up in comments. Nothing is too weird for me. Really. Nothing.*

[7] http://io9.com/tag/fully-functional

Jens Ohlig

DATAMINING SLASH FICTION WITH AUTOMATIC TEXT ANALYSIS

The topic of homoerotic amateur science fiction narratives, or slash fiction, has been the subject of several academic studies. Originally a part of the fan culture around the *Star Trek* TV series, the name comes from the slash symbol (/) in the description of the primary pairing involved in the story, such as 'Kirk/Spock'. While most works in academia look at the phenomenon from a cultural studies approach and discuss questions of gender trouble in pop culture before the appearance of openly gay or bi-sexual characters (such as Willow and Tara in the television series *Buffy the Vampire Slayer*), we feel that automated and computerized analysis of the works as text may bring in a new perspective.

Slash fiction is a phenomenon in sci-fi fandom that deals with stories about romantic and sexual relationships between characters of the same sex, originally in the *Star Trek* series which first appeared on TV in 1966. Interestingly enough, slash fiction is almost always produced by and for heterosexual women. The quality of slash fiction stories differs wildly. There are pieces of higher literary quality as well as stories that cannot be called anything but a few lines of smut. Stories can range from romance to pure pornography. While classic fan fiction focuses on exploring technology and social rules of a given SF universe, slash fiction emphasizes only point: relationships between same-sex characters. Ties to sci-fi fandom and general fan fiction are there, but they are much weaker than one would expect; to a large extend, the slash fiction community can even be considered outside the sci-fi fandom community proper.

After having started in the late 1960s, slash fiction moved to the Web in the 1990s. This step was almost Vulcan in logic: by definition, slash fiction can only exist as copyright infringement, almost never to be published by conventional publishing companies, and thus is an example of illegal art. For our analysis we were able to benefit from the fact that due to copyright restrictions very little fan fiction is available in print form and thus newer works of the genre are published almost exclusively on the Internet. This means that a digital text corpus for computer analysis is readily available.

Relatively vague literary criticism terms like *motif*, *theme* and *myth* can assist in classifying this corpus. But as computer nerds, we took a different path: writing code to deal with the problem. Our model is that of social network analysis (SNA). We looked at the texts not as a collection of words or literary works, but rather as a network, a weakly connected simple digraph in which every arc has been assigned a non-negative integer, the capacity of the arc. For our research, the arcs of our slash fiction network are the textual elements (lexemes, phrases), while the capacity represents the value of the textual or topical relationship of these elements. Using network analysis and topic maps, we dug into what constitutes slash fiction statistically and semantically and visualized the data found on some of the more interesting structural elements in the text corpus. Through SNA, we tried to detect the relationships between and among words and themes for an analysis of quantitative characteristics of textual data as well as the extraction of meaning from texts. Most of the software used for the automated analysis and data visualization was written in the Java programming language or a variant called Processing specifically for this research project. For us, concepts in the stories were equivalent to nodes in a network. A link between two concepts is a statement, which corresponds with an edge in SNA. Last, but not least, the union of all statements per texts forms a map.

The process of data visualization typically consists of 7 steps: 1. Acquire, 2. Parse, 3. Filter, 4. Mine, 5. Represent, 6. Refine, and 7. Interact. Our initial goal was to acquire the definitive corpus of slash fiction; however we had to give up on that. It became clear that a comprehensive corpus is probably impossible to collect. Our approach for acquisition was thus to try out our visualization on randomly picked stories. We decided to concentrate on stories with the pairing of Kirk/Spock.

A sample map of a slash fiction story represented as a network can look like this:

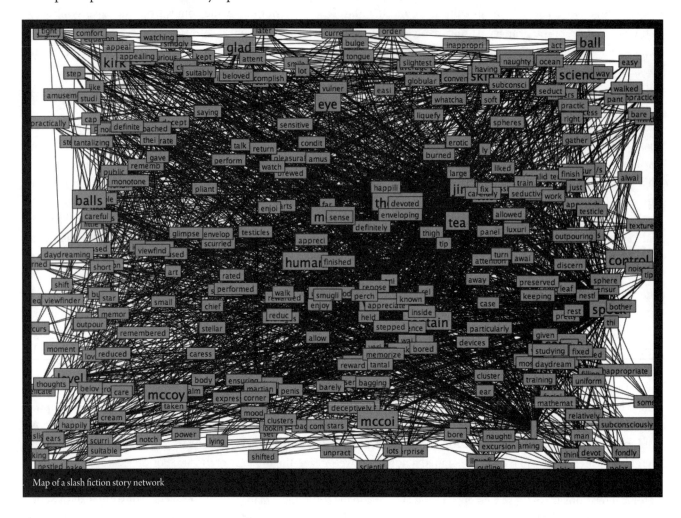

Map of a slash fiction story network

The parsing process for that is pretty straight foward:

```
/**
* Loads the data from the corpus
*/
void loadData() {
String lines[] = loadStrings('stargazing.txt');
Stack tokens = new Stack();
for (int i=0; i<= lines.length-1; i++) {
  String s = lines[i];
  // Read in the tokens, convert to lower case, and push them on the stack
  String[] p = splitTokens(s, '-\",.!?\r\n; ');
  for (int j=0; j < p.length-1; j++) {
   p[j] = p[j].toLowerCase();
```

```
    String token = p[j];
    tokens.push(token);
  }
 }
 addStackToEdges(tokens, defaultWindowSize);
 }
```

Filtering is the process where we check against a list of stop words. For a text written in English, typical stop words are words like 'I', 'a', 'the' etc. These words are seen as too insignificant to include in a network of concepts, so they need to be filtered out:

```
for (int j=1; j <= windowsize; j++) {
  if (i+j <= t.length-1) {
   for (int k=0; k <= stopwords.length-1; k++) {
    // Check for stopwords
    isStopword = (t[i+j].equals(stopwords[k].toLowerCase()) || t[i].equals(stopwords[k].toLowerCase()));
    if (t[i].length() == 1 || t[i+j].length() == 1) {
     isStopword = true;
    }
    if (isStopword == true) {
     break;
    }
   }
   if (isStopword == false) {
    addEdge(t[i], t[i+j]);
   }
  }
 }
```

Refining the data could mean to further reduce the nodes in the network to make the concepts and how they are related clearer. One possible idea is stemming, i.e. reducing words to their stem, cutting off '-*ing*' or plural forms. This approach, however, doesn't make the network clearer. Due to the nature of the texts examined, it is e.g. desirable to leave 'balls' in the plural form.

A more interesting approach would be to look for synonyms and holonyms, especially for body parts and examine the resulting network, when for instance the nodes 'cock' and 'dick' are reduced to the concept of 'penis'. For this, standard computer linguistics libraries like WordNet may come in handy.

The approach of representing words in a text as node in a network may not provide us with a story grammar for slash fiction from where new stories could be automatically generated, however we think that our visualization of what is related and connected in amateur homoerotic science-fiction stories may at least lead us to look at these pieces of literature in a new way. After all, it's all about going boldly where no one has gone before.

Panel with Violet Blue, Benjamin Cowden, Daniel Fabry, Stephane Perrin (23N!), Allen Stein / Host: Johannes Grenzfurthne

THE EROTIC OF THE MACHINE
ARSE ELEKTRONIKA PANEL 2008

Johannes: Welcome to our big panel. It's quite gender-imbalanced, I have to say. All of the wonderful women we invited today to come on stage cancelled, just like an hour ago. But I promise we will try to genderbend ourselves, right here on stage.

Our topic is 'The Erotic of the Machine' and I would like to introduce our panellists. Here we have Daniel Fabry; he's part of monochrom and co-editor of our 'pr0nnovation' book. Daniel teaches information design at the University of Applied Sciences in Graz. He deals with user interfaces a lot.

We have Stephane Perrin, who is from France but actually at the moment lives and works in Japan. He's one of the winners of the *Prix Arse Elektronika* this year, with his wonderful series of controversial dildos. And I think his work also concerns death and sex and erotica of the machine. It's wonderful to have him on stage tonight.

We have Benjamin Cowden, whose machine 'Eating My Cake and Having It Too' also won a prize. As I said, it's a piece about outsourcing the fetish to the machine, so we don't even have to do it ourselves.

Please welcome Allen Stein of thethrillhammer. I have to say: he's one of the big men concerning penetration devices. [laughter]

Allen: Teledildonics.

Johannes: Teledildonics. That sounds nicer, yes.

We have Violet Blue. She's nearly everywhere. I don't know where she is not. She's a sex blogger and sex educator?

Violet: I thought you said 'agitator.'
[laughter]

Johannes: Well, even that. She's just fabulous, and it's wonderful to have her here today.

And we are very happy to be able to welcome Thomas Roche. He writes erotic fiction, is a sex educator and he's very interested in BDSM.

Thomas, what is erotic about a machine?

Thomas: For me, there are a couple different aspects to that. I think that nowadays when we think of machines, the word 'machine' conjures up, for me, something that's more like a factory or like kind of a pumping piston almost steam engine sort of thing.

But I think that a machine is any mechanical device that does what we tell it, or almost what we tell it to, or argues with us. So a computer is a machine. And just speaking for myself, my sexuality has been completely enabled by computers. I think that's true of most people.

Many, many people with any kind of divergent sexuality or anything that's esoteric about what they're interested in can now get much more of what they're looking for by using computers. I think the field of sexuality has been completely changed by what are probably the most important machines available now - which are information machines.

We have to remember when we're thinking about machines and what are erotic about them, there are two things: there's the physical and the innovation. That said, I also think it's important to look at the ways in which machinery kind of defines human development. All of the movement of people moving into cities in the 20th century has largely been enabled by the fact that we have easy, cheap machinery that makes those things possible. It makes transportation possible; it makes all sorts of relocation and shipping of goods possible.

Machines are such an integral part of our life that it really defines what our social world is like. And machines themselves are just really, really hot. Definitely I think there a huge sexuality to them.

Lester Bangs, who is a great rock and roll writer, said 'What is the difference between the curve of a breast on a sex goddess and the bones in the thighs of a stud and the fins on a '57 Chevy?' I'm with Bangs; beauty is beauty. Any work of art evokes the same feelings of appreciation, therefore conceivably eroticism, in humans. What is machinery if not a type of art? Anybody who builds or invents things knows that it's got art in it as much as science, and a lot of improvisation. Machinery is art that works; it's art that does something. It may be practical, but it's an art form of getting it to that point, where it's doing what it practically is supposed to do.

So I think all aspects of human experience are reflected in machines. That's my perspective.

Johannes: That's a pretty industrial perspective, I guess.

Thomas: I guess it is.

Johannes: When I think of machines I think of the 19th century.

Thomas: Even the word has sort of this connotation of industrial like pumping sort of machinery and all of those things are very important. But a machine is anything inanimate or presumably inanimate that interacts with us that we are able to task to do something.

And the things that we task machines to do are increasingly intimate. Even 10 years ago, the variety of sex toys that were available was minimal, at best. And now it's amazing the variety we have. Let alone that now you actually have machines that will provide thrusting motions, so you can actually get off if you're looking for a more complex or varied interaction with machines than, say, vibration, which is all that was available a decade or two ago. Just the things that we're expecting machines to do are increasingly parts of our lives in ways that they really weren't in recent memory - even ten or even five years ago.

Johannes: I think it would be a good idea to pass the microphone to Violet. There's this Cambrian explosion of sex toys in the last 10 years. What do you think about that? What does it tell us about society?

Violet: In general I think it tells us that society is becoming a lot more tolerant of the sexualization of machines and robots in general. But I remember seven or eight years ago when I was working the floor at Good Vibrations, when I would be selling vibrators and male customers would come in to buy a vibrator for their girlfriend for the first time or for their wife for the first time.

They would often joke, and it was often something that we would have to sort of smooth over with them, the notion that they would not be replaced by this machine. Because it was a constant insecurity that was being brought up, that this machine was something that could replace them in the bedroom in some way.

And I think that speaks a lot to what is erotic about the machine too, and why the interest is growing and growing. It's not just mystery as to why someone would want to fuck a machine, but also there's a certain power exchange that goes on between the human and the machine and the hardness of the machine, and the softness and the vulnerability of the body. I think it's just utterly fascinating.

It's something that has been going on for a very long time. I'm going to totally massacre his name, but Villiers, wrote 'The Future Eve' in 1886. It was the first human-android sexualized relationship written in novel form, about a man who cheated on his would-be spouse with an android. So we have this long term fascination with making love to the machine and having sex with the machine. In that aspect of the men having sex with the female gynoid which translates just as easy today.

David Levy wrote 'Love and Sex with Robots'. He's predicted that marriages between humans and robots will be legal within five to 20 years. But at the same time, we have people using online services to marry robots and artificial identities online, and making sex dolls. So we have the 'Future Eve' suddenly here.

Johannes Grenzfurthner: We should probably declare gay people robots. So at least in 15 or 20 years they will be able to get married.

Violet: Well, Levy did write a piece about the rights of robots and the rights of artificial intelligence entities - which is really interesting when you think about what artificial intelligence developers are doing.

A lot of activity is happening with the virtual girl, the virtual date, the KariGirls. There's a really thriving online community of these men who have these artificial intelligence girlfriends. They're based on the A.L.I.C.E. model which is Dr. Richard Wallace's chatbot user interface. They're developing them and they're having these ongoing relationships with these women that they're creating. They're marrying them in these online chapels and having long term relationships and sexual relationships with them. We're just this close to having them put into a RealDoll body, I think. It's fascinating. It undermines a lot of our assumptions about sex with robots.

Johannes: Allen, it's almost a philosophical question, but what's really the difference between a machine and a body? I mean, a body is a chemical machine.

Allen: Beyond the biological and emotional debate of man vs. machine and the biggest difference is that there's no energy exchange with having sex with a machine. Machines are never going to replace human interaction. It might be a nice accoutrement to the whole thing, but there's no energy exchange on the energetic level. I suppose a tantra-based machine might be kind of cool but it will be a while until thought, emotion, and the self-awareness of machines take place.

Violet: When you think about the Japanese Tamagotchi craze - there was this one woman who had an utter meltdown. This is a little mobile creature that you carry around, a key chain device, and you 'feed' it with 'love'. There was a story of one woman who had an utter meltdown when she was told that she couldn't take it on an airplane and refused to fly because she didn't want it to die.

Also, there was some questioning of Hebrew law in Israel law about shutting off devices, and there was a lot of discussion about the fact that kids were going, 'We can't shut these off because they'll die.' There are some questions to be raised about emotional bonding with machines, definitely.

Allen: Emotionally, true. People can become attached to the strangest things. A fine example of that is the touching movie Lars and the Real Doll. Energetically, I just don't see it on that side. It is interesting regarding the Hebrew law. I think letting the tamagotchi die would not be kosher.

Johannes: Tell us about what you're working on at the moment, what are your projects?

Allen: I have a couple of projects I've been working on lately. The first is teledildonically based - sexmachinecams.com. I'm setting up a studio in LA currently, that has about 15 Internet-controlled sex machines from Sybians to basic stroking machines. There's a guy in New Orleans that makes these well built machines, Ken's Twisted Mind and I have added my boards to them. We also have the new thethrillhammer down there as well. People can go online and remotely pleasure a male or female model from their own personal computer.

Besides coming back out and doing some more teledildonic stuff, I still shoot sex machine porn to fuel my research into actually doing some biosensor work and developing wellness products. We are developing a suite of products to help society to have better sex through gathering their biometric data and translating that data into a comprehensive sexual wellness program - better living through sexual technology.

Johannes: We have Thomas Roche who's working for kink.com. Is there a battle between people who are doing fucking machines porn? The scene, I guess, is not too big, but is it competitive?

Allen: I don't see it as a competition at all. It's been very strange, because when Kink started launching fucking machines I started doing Internet controlled sex machines. So their business model was women on video, whereas mine was a more live, one-on-one videoconferencing play. This is eight years ago, so we're looking at Java push from Windows Media, where you interact with a machine, you have to wait five seconds to see it catch up to you. That thrust me into the content game and for awhile there was really only a handful of us shooting machine porn. Even in the sex machine scene there are many different niches there. My porn is less BDSM kink orientated so the stuff I produce appeals to a different crowd. I love being able to say, 'Hey I'm not a real doctor, but I play one on the Internet!'

I think it would be great to work with Kink, especially with their arsenal of machines. I'll just have a heyday hooking up my internet controls to them all and being able to bring a live aspect to it.

Because what I'm seeing on a commercial standpoint with teledildonics, is that viewers come into the shows with these chat models, they're gonna do webchat anyway. But when they come in there, and they're able to control the machine and actually have consummated sex, they start forging a relationship with that performer. And instead of being a customer, that customer turns into a regular, comes back more often because they are interacting on a level that wasn't there before.

Thomas: I'd love to say something about that. I don't see any of it as competition as such. In a sense I'm not on the business side of the business, I do public relations so I want to be everybody's friend. But I think that all businesses that explore this stuff are really doing something that's a net positive.

The interesting thing from that perspective is that anybody who is doing sex machine development is developing equipment that may be doing the same basic set of things, but they all do it differently. The variation is so extreme. Especially in a field that's so overloaded with robotic, for lack of a better word, human-on-human interactions that are sort of predictable and the same sort of thing and all the models look more or less the same. And even if there is more variation in body types and styles of models, they're all human.

The machines actually offer a whole new set of things that I think is just barely being explored. I mean, the variety is enormous, because these things haven't been specifically eroticized before about ten years ago, roughly. And I know that when I was growing up a complaint that I often had, was that vibrators will be shit. Now vibrators rotate, dildos rotate. So when I started seeing thrusting sex machines that offered more variety... I knew something was changing profoundly.

Allen: Machines are getting more and more sophisticated as the motion control folks come out of the woodwork with more feature laden machines. Really the sky's the limit when it comes to designing machines. It is all usually held back by budget alone. As they say, 'Can you really put a price on pleasure?'

As far as access to technology, there are three people, actually two now, my company and Highjoy Products that do commercial teledildonics and many manufacturers doing their own thing. Bottom line, controlling sex devices on the Internet is simple, you're just controlling voltage. So it's really not that hard to do. But getting the different service providers to work together and be able to have someone who has a Highjoy device and someone who has a thethrillhammer or a Sybian that's Internet controlled, being able to have sex with each other someone has to say, 'OK, I have this control, and I have this machine over here that we need to control let's pop the controls for that machine.' So what I'm working on now is more of a teledildonic corset of APIs, that people can just grab the widget tied to a basic core API so they can go fuck each other easily.

Johannes: Benjamin, you're coming from a completely different field. You're coming from arts. Your machine wasn't constructed to be a sex machine. Did you think about the erotic dimension? What's your opinion on the erotica of the machine?

Benjamin: I guess I tend to be a lot more classic and a lot more art focused in my viewpoint and I'm thinking more along the lines of something Violet mentioned; that the late nineteenth century, early twentieth century, the tail end of the industrial revolution was when we started as a society in the west falling back in love with the machine. The beginning of the Industrial Revolution was so dark, everybody was afraid of machines. Later, as the benefits of an industrialized society became more noticeable, machines were looked at more favourably, and that's when artists stared utilizing the image of the machine as erotic.

In a way, I feel like it's a 'god fantasy' on our part, to envision a machine as erotic. Because we create the machine in the same way that we in the West tend to imagine that god created us. The machine acts independently from us the same way that we act independently.

I think this issue hinges on Western mythology.

Johannes: But you could also say, especially if you're talking about Taylorism and Freudianism, that people also start to create robots out of people. In assembly lines people are slaves to machines. Are people falling in love with their assembly line?

Violet: It's a power exchange.

Benjamin: Yes. Throughout the 19th century people in general were afraid of machines and thought machines were bad. I'm talking about lay people. Even though the rest of the industry because of the economy was moving forward no matter what people thought. There was a generally negative view of the machine.

But once the Industrial Revolution hit its stride and quality of life became better, at least economically, from the progress that machines were making, people saw things differently. It sounds weird to praise the Industrial Revolution. But people started having a more positive view of the machines, and that is when people started being able to look at machines as erotic.

Johannes: You were mentioning the Western mythology and Stephane is from France, but now he is living and working in Japan. It would be interesting to hear about the Japanese perspective. Japanese pop culture is soaked with machines and eroticism.

Stephane: When I learnt about this conference and started to think about what I could do, I looked at what had already been done for these kinds of things.

In Japan it's kind of not very popular for these kinds of things to be so big. You have no room to do these kinds of things. So I think that usually when you think about something there, it's very rarely a mechanical thing - a very big thing.

So it was more like just working on my computer.

Johannes: Allen, do you have any customers in Japan?

Allen: My customers, I had more data on that when I was running membership sites, so I could see where the traffic was coming from. I did have traffic from Japan, but it wasn't that significant.

What I do is I produce sex machine content for a variety of other companies. So each company has their own niche, and each little niche has their own version of what's erotic to the niche.

For example, I shoot for one company, Homegrown Video, and they want amateur women. So it's amateur girls on machines having real orgasms. And the draw to the series itself, that niche, is real orgasms.

Whereas for another one, Wild Fuck Toys, I play a doctor and I cure people if their orgasmic ailments on a variety of these machines. That's more appealing to a different type I would say the 'Maxim' type of crowd. Like people who say, 'Dude' a lot.

In Japan I would assume it would vary culturally regarding the eroticism of machines. So culturally it varies depending on which specific niche the machines are in, I would imagine.

Violet: I wanted to add something to that too. It's interesting you bring up the efficiency aspect of the Japanese take on sex and machines, because one thing I've learned in working in the sex industry and the sex toy industry is that some of the best vibrators in the world come from Japan.

They've really, really refined the art of the vibrator there, and they have for a very long time. Japanese vibrators are very expensive, they're made of high quality materials, and they have excellent motors. They come in mindboggling array of shapes and sizes. Some are very scary looking.

And they've done a lot of really interesting development around doing different types of add-ons for sex machines. I just recently blogged about the Eroy Onacup Blowjob Station Machine: it looks like Japanese little girls' school box or lunch box. It's really creepy looking. It looks like Hello Kitty. But it's an attachment for a fleshlight, which is a male masturbation sleeve device. And basically what it does is it turns it into a thrusting device. So it looks like a little lunchbox gun, essentially. It's very interesting.

At least in my history of working in the industry, the best sex toys have always come from Japan. The weirdest sex toys have always come from Japan. They just seem to keep developing and keep developing and pushing out more. It's really fascinating.

And there's a lot of other mystique around sex toys from Japan, and stories about why they're not physiologically shaped and things like that. Did you have something to add?

Thomas: Yeah, I just wanted to add very quickly that for what it's worth, one of the very few times Kink.com, where I currently work, has gotten into DVD production has in conjunction with a Japanese company that releases fucking machine DVDs. So I just find that interesting that there was that interest. And they approached Kink. And I think it does have appeal in Japan in that way.

About the video of the guys packing up the machines to go to Japan, it was priceless.

[laughter]

Stephane: I just wanted to share something more about Japan. They have a highly advanced technology for dolls, as well as a market for reviews and books about these dolls. I find them pretty creepy because they are very well made and extremely expensive. There are a lot of companies that specialize in making them. They kind of do case by case models, so you can ask for whatever.

People sometimes ask for strange things. And a lot of these reviews contain people acting with these custom-made dolls, combined with pictures - or explaining methods of changing the elements on the dolls. I've even come across several pages in the midst of one of these books just all about the vagina, combined with statistics.

It's quite interesting but sometimes it can be a bit strange, yeah.

Johannes: So, Daniel, your specialty, and you've been working on this for a long, long time, is the Theremin and other touchless devices. I would be interested in what's your take on the erotic of the machine?

Daniel: I think the Theremin is maybe a good example for an interface that is more erotic than I don't know a button or a keyboard a mouse. Because it could leave a little more space for fantasy, which in my opinion is a romantic feature of eroticism.

It makes more open space to your fantasy, your connections, your experiences. You could connect a chest approaching to an object or just enter a room - so you don't touch it but you have to get closer to it. I think that's just more interesting than touching a haptic device.

Thomas: So would it be interesting to think about not only like fucking machines, but whole fucking spaces? In a certain way, that you're moving around in a space and you have a whole environment, like a central sexual environment that's not only like, you know...

Daniel: Oh yes, absolutely. That's the whole cyberspace and cybersex thing, which was a big topic back in the eighties and nineties, but there was a drop off. Now it's coming back slowly again - all the multi-display environments or immersive environments in exhibitions, for example, where you are sensed by sensors and have feedback by audio, video, et cetera - which is very immersive if you have surround projection. So it's all possible - it could be an interesting topic to make erotic rooms and use the Theremin as a sensor...

Johannes: Everyone was thinking of the bright future of virtual realities, but it never really happened. Wouldn't it be better to think about augmented sex realities? I mean, having like...

Daniel: Eyeglasses. Walking through the street...

Johannes: Ha, like really cheap ads in those 1950s pulp magazines, where they tried to sell you 'x-ray glasses' to see women naked. Remember that?

Daniel: [laughing] Yeah. I wonder if anyone really bought that. I don't know...

Johannes: I'd like to throw the question in the round regarding augmented sexual devices - any ideas?

Violet: It's sort of interesting, while you're talking about the way a Theremin or the way that the touch and the movement of working a Theremin could work itself into a sexual environment sort of was making me think about what you were saying earlier, Thomas, about the intimacy of our gadgets. And so I was brainstorming a little bit about, 'What was the consumer side of something like that look like?'

There's something called a VGirl that you can download and she is a virtual girlfriend. She's an artificial intelligence entity and there are five or six different types, they're different skin tones, body types, etc., etc., and of course there's an ongoing fee to have this relationship with the VGirl. You get her to undress and do sexy dances for you and things like that, and she's constantly sort of leading you on to, you know, continue the payments.

But what I'm wondering is how this is going to tie into the purchaser, because I pay attention a lot to cell phones and the cell phone market because it's one of those devices that is just always one of the most intimate devices we have, we have it with us at all times. With a touch screen, I wonder what the possibilities are going to be with computer controlled, you know, doing teledildonics with your iPhone, you know? Like fucking a girl with your iPhone while you're on your commute, or something, or...

Johannes: I'm not a good programmer, but I often think about a device that pretty much works like licking an iPhone.

Thomas: I just wanted to say, I think that's a really good point, that we forget that there are two things: there's input and there's output. And the talk that Kyle Machulis gave last year was largely about output devices, about things that can stimulate you but the input is equally important to a real in person sexual enterprise.

If you were just licking a touch screen, that would be an example of an input device that wouldn't have been possible in earlier generations of computers, at least not at a consumer level. Now, licking a touch screen may provide a very funny image but at the same time I think that it is important that we have these touch screens and we're increasingly interactive with input devices that react based on what the software state is - the state of the device is and how we're physically touching it. So it's very, very sensitive and will be increasingly sensitive to the kind of finer points of human touch that maybe can be rendered digitally, but nonetheless don't feel digital, they feel organic. The inputs that the human body can provide in a sexual situation really are very complicated, and they're extremely subtle.

I wonder if the place that we're going to see this really sort of intimate development isn't going to be in the fine arts, where computer artists are creating all sorts of things? I'm not an artist, but I know that the artists that I see work with a few devices that look a little more complicated than my mouse or my touch pad.

Think of visual artists and the subtleties that they need to be able to create a piece of visual art on the computer being similar to the kind of work that you need to do in the physical space to have a good sexual experience.

Johannes: Daniel, is there anything going on in the realms of art and exhibition designs?

Daniel: All the interaction designers I know are exploring different interfaces now and different interface technologies are becoming more a cultural technique, which you could say like this.

If you have an iPod, for example, it was a slow development from buttons to a wheel and the next step just the wheel without buttons - it took some time to make it as a cultural technique, also touch screens, for example.

So there are really, really a lot of interfaces and technologies now on the very high state – they are accessible and also affordable. But I think the tricky thing about that is to make the right content and to make the right interaction design. For example, the Wii controller - the technology, we had it in the seventies or eighties. There were several tries for a kind of different interface and lots of them crashed on the market but now the Wii somehow made it and I think it doesn't take too much time to learn a very broad audience or users how to deal with the Wii.

So it's always the same - exploring new interfaces, exploring new technologies, but the way you should be thinking is: what's the right content and what's the right interface for it. And you design for it.

Johannes: Stephane, what's your take on interfaces? I mean, you're dealing with interfaces a lot.

Stephane: Yeah, I like to work on the cell level, all different sorts of things and measurement devices. I haven't seen a lot of work done about how to create erotic sensations beyond penetration. It's like if you have a sensor, you can tell the temperature or humidity of a body - if people are sweaty or excited. They never talk about possible sexual applications based on such measurements. So I think it's a kind of interesting new way to have more sensors about body states involved. They could be very small.

Johannes: In art, especially in the 1960s and the 1970s, there was a big focus on body art, especially in films by David Cronenberg. His films are dealing a lot with environments and embedded devices. There is a wonderful scene in Videodrome where the guy pretty much crawls into the TV screen is such a brilliant erotic and uncanny way.
Benjamin, could you give us examples of really interesting art projects of the last 15 years that dealt with the erotic of the machine?

Benjamin: Well, I haven't seen a whole lot going on in the art world currently that I find extremely interesting in term of eroticism and the machine. I feel like machines in general are pretty unpopular with artists right now.
I've been thinking about what we've been talking about, and we have mentioned touch screens, and we have mentioned the Theremin. I find the Theremin to be immediately erotic. But I don't find touch screens to be erotic. They may be a medium for erotic content. In terms of the inherent eroticism of a machine, I have trouble thinking of other things that are actually popular right now and not confined to small social subgroups

Daniel: It just came to my mind that I did an installation using the Theremin for controlling some kind of erotic content. I don't know the movie I used, but maybe that's not too important. It was a movie a popular with a lot of brutal content and sexual content, but basically it was about the relationship of a girl and a man.
I took just a couple of scenes where the characters are touching each other or having sex with each other. And so the scenes just switched from one scene to the next and you could approach an interface which was a white mask, a strong symbol in this movie.
You could approach the mask and control the movie clips in its timeline, for example, the couple having sex or another scene where they hit each other. It was displayed at Ars Electronica in Austria.

Johannes: The one in Linz?

Daniel: Yes, in Linz. And people were really afraid of using it because it somehow involved them in something that they saw in the movie clips. But they realized that when they came closer, they were somehow controlling the actions in the scenes - having sex or hitting each other.

Johannes: They didn't want to?

Daniel: No. Some of them really enjoyed it, but others said 'What am I doing here? I'm having sex or I'm controlling people having sex with each other,' et cetera.

Johannes: So probably that's something not for the public sphere - maybe a business model for home technology.

Thomas, we've been talking about interfaces. You're part of kink.com. Is there anything interesting going on concerning interfaces at kink, com?

Thomas: We haven't done a whole lot of interface work. Generally, the model the woman who's getting fucked is controlling the device, or someone off-screen or semi-off-screen is controlling it with the same sort of dial or button or the like. Essentially, they're fairly primitive inputs, though they might provide a complicated result through the way they go together.

The important developments in input, as far as sex machines go, were sort of made early, I think. Most of what is being developed in terms of sexuality is more around the social networking area. Which I think will become important in terms of online one-to-one interactions with machines, but I don't think that interface work has really been done yet in relation to sex machines.

I want to say one thing about social networking and sex machines, though - one of my earliest jobs in the adult industry was working for iFriends, which marketed a cyberdildonic or teledildonic device that allowed someone in an iFriends chatroom to interact on their computer with a model who was on webcam. And this was before USBs were available, so the way the technology worked is that the user would stick a sensor on his or her monitor and then call up an on-screen interface - and the changing colors would control the vibrator on the other end, and the girl would ooh and aah, but I'm sure it wasn't very satisfying for the recipient. This was just about ten years ago. In terms of broadly-used interface technology for having remote sex, things haven't changed that much. Most people who are having remote sex with a lover are pretty much getting on a headset and having someone on the other end say 'Yeah, put your hand on your dick,' and they put their hand on their dick, or 'Turn up the vibrator,' and they turn up the vibrator. There's technology to do more than that, sort of, but it's not broadly available or broadly used by the general public on a peer-to-peer or peer-to-sex-worker basis.

Allen: We offer full interaction with our sex devices now over the internet at sexmachinecams.com You are going to see a proliferation of internet enabled sex devices coming out in the next couple of years as chips make them into sex toy manufacturer's designs. In the design of our interfaces, when we started getting into machines controlled on the Internet, the first thing we were looking at is what were other people doing? And one of things out there was this nightmare interface for the Sinulator. It had this dashboard and it looked like a video game. And that is so far from being erotic to anybody.

Johannes: Kyle presented it last year as the worst possible sexual interface out there.

Allen: It was awful. You've got to walk a fine line, because you really want to design an interface that's intimate but still have the usability factor there. Because you've got to figure these people who are going to be controlling these machines are busy at the moment. It's got to be as simple as possible and we have to make it intimate at the same time. So my controls are very, very simple.

Johannes: Cronenberg's interface in eXistenZ is so wonderful - because it's soft and flexible. Is there anything going on concerning soft interfaces?

Thomas: I wanted to say one thing about that. I think eXistenZ is a very good example. It's really interesting to think about that. Because in this whole Cronenberg thing is this organic machine, not so much that humans interface with machines but we sort of machines are built out of human tissue or some sort of biotissue. Which I think is interesting.

But as far as sex machines go, the first time I ever used a sex machine, it was very simple it was just a dial that made it go fast and it was very, very sensitive. I was blown away. This was probably about 2000, maybe 2001.

I was completely liberated by the fact that it was simple. Because for me, having actual in person sex, the complications of all the physical stuff is actually more confusing to me than dealing with digital data, since I'm used to dealing with words, primarily.

So I actually find it very appealing, the idea that this is a very simple set of axis, and it actually helps me envision non-machine, real world and entirely biological sex. To think of it in terms of a grid or a set of stimuli rather than this very complicated thing that you might see in Cronenberg. Which is what real interfaces are like, real biological interfaces are like.

Violet: I think the closest we're going to start to see the widespread use of something like that is through companies like RealDoll, who are actually making large, soft, sex toys that are very popular and very available to consumers who have money. They're around $6000.

And they've been working on putting actuators in their hips and motorizing them for quite a while now.

Thomas: There was a dental robot that was designed to train dentists in Japan, and which the robot would respond if when you were working on her teeth. It was a female robot and it looked very realistic. It was very strange.

Violet: If you touched her breasts, she would tell you to stop.
[laughter]

Thomas: Right, this is very interesting. She would react if you hit the wrong nerve. Like if you drilled and did it wrong, she would react in pain which obviously is...

Johannes: It's puppet empowerment.

Violet: Thomas, we're talking about Fucking Machines, and Fucking Machines has been a really popular site for a very long time, and it's probably made piles of money.

Basically, what it is, it's just porn. It's girls fucking machines, girls being fucked by machines. Why do people want to see that?

Thomas: Why is it so popular? I've heard there are two competing theories. The theory that I hear most often is because it's heterosexual men who don't want to see other guys on camera, but find it hot to have girls get fucked.

I actually don't buy into that. I believe there's more to it than that. I think that the fascination that women have sex with the machines, I think, is because they fantasize themselves being serviced in that way. I think that's really hot to a lot of women. But even just the visual pleasure that women take from it, I think, is really intense and I think there's way more to it. I don't think it is just people who don't want to see guys. I think that it's just all sorts of the variety that's available from a girl being fucked by a machine.

Let's face it. We can build a machine to fuck a girl a lot harder, and a lot longer, and a lot more steadily, than a human is ever going to be able to do. I think that's a lot of the appeal.

Violet: Thomas… in fact, Kink.com, actually, you set the world record for female ejaculation with some of your fucking machines in a measured competition. I think it was 15 feet. Was that...

Thomas: It was Annie Cruz. She's kind of a special case, because when you've got her, she's just like a copious ejaculator. It's kind of a special case. That's almost cheating, because it's almost like having Superman doing weightlifting. [laughter]

Johannes: In our pr0nnovation book there is a one page short article by Binx, who had sex with Fuckzilla last year at *Arse Elektronika*. She describes it as an act of third wave feminism.

Daniel: It's a different case because people are controlling the machine. They're not watching women and the machine, per se.

Allen: Per se, yes. Why do people want to see machine sex? Well, the porn I do shoot is for a couple distinct markets. So, the stuff I shoot in the amateur market is for people who want to see real orgasms. It's about real women and real orgasms for that series and that audience finds that highly erotic. That's what got me into Internet controlling machines. Real orgasms.

Then, there's the Maxim crowd, or the 'dude' crowd, who like to see women just get fucked by machines. It's less erotic but appealing nonetheless. John Henry may of beat the machine but he dies in the end, right?

Then you have the control aspect of the online teledildonics, where they actually get to make the relationship. They can 'control' the machine. They can extend themselves through a sense of touch and physically affect the other person. That provides a great sense of control for the driver and a lack of control for the rider. Either way that translates from a commercial standpoint into longer shows.

When we do a price elasticity study on it, yes, we could charge $11 a minute for it. But if we charge a normal rate and just do it as added value, people stay in there twice as long.

I found that some people are interested in just playing with the camera, and not even driving the machine. Many men are there for the company.

Johannes: Is there gay machine porn?

Thomas: Well, Kink did a site called Butt Machine Boys some years ago. It was a limited success, for a variety of reasons. The argument that I've heard as to why that didn't really work is because guys who want to watch guys get fucked, want to watch guys get fucked by other guys. I don't know if I necessarily buy that.

I'm not really sure why that didn't work. That's the only gay machine porn in large scale that I've ever encountered. But it must exist somewhere. Do you know, Allen?

Allen: No.

Allen: Not quite sure, but I placed my initial ads for hiring for our studio in LA for sexmachinestaffing.com and it's been overwhelmingly 70 percent men, 30 percent women, from the general populace. And the men want to come in and get fucked all day.

Johannes: OK. We're here in the Bay Area. We're here in the center of techno-culture. But is the web industry really ready for the erotic of the machine?

Violet: I think we're the epicenter of it. We've been the epicenter of so many aspects of sex culture, in general, and tech culture. I think there's a really good reason for that, and I think it has to do with the types of people that are interested in technology and the types of people that are interested in sex.

I think there's a lot in common with both types of people. I mean, there's so much overlap when you come to a conference like this, or you go to a hacker conference, or you go to a sex conference, you often run into a lot of the same people.

I think that, it's not that we're ready for it, I think that we've been doing it all along.

Johannes: But many people I talk to in the Bay Area are still kind of offended by the whole topic of our conference.

Violet: It depends on what circles you end up travelling in, around the Bay Area.

Johannes: Google?

Thomas: It's also just about money. Silicon Valley has a large number of people with a fair amount of income. They're the ones who can afford to experiment with all sorts of stuff, down to buying an iPhone, even if you want to just get really basic about experimentation.

Violet: It reminds me of Metropolis, where the idle rich would have their children play in the gardens above, while the people toiled with the machines below.

Johannes: There's this long back story about destructive machine art in the Bay Area. Mark Pauline and many others constructing big machines killing each other. Rather Freudian, isn't it?

Thomas: Let's get Freudian.

Johannes: Caution. Freudian slippery when wet.

Thomas: I think it's fair to say that we make machines, so we're going to make them sexy, right? Don't we do that?
I think anybody who's involved in engineering, even product design, engineers, and artists well, let me clarify. Artists that make kinetic and interactive work want the things that they build to make people want to touch them.
And that's what I think about. I think that's a common thread. I even think Mark Pauline and Survival Research Labs I mean, those machines aren't the kind of things you're going to want to touch, but the implied power in those machines is kind of erotic, in a way. It's destructive, but that calls to mind the whole idea of creation and destruction.
There's something sexual in that.

Violet: I just want to also add that, if you take a look at Benjamin's work online it's twentysevengears.com … To me, it reminds me of an artist named Arthur Ganson. You use a lot of worm gears. They're very sensuous. If you take a look at that machine over there, and the way it works, and you look at his work, a worm gear, the way it's shaped, it's very curvaceous, it's very tactile looking and very tactile feeling.
So it's almost impossible not to want to sexualize something like a worm gear. Just simply by the way it moves and it undulates when you turn it with your hand. And when you talk about machines at Survival Research Laboratories, which I worked on for 12 years, they're dripping with grease and they're dirty and they're powerful and they're loud and they're overpowering. They're insanely sexy for all of those reasons.

Because it's just the way of watching the gears together. It's sloppy and it's scary. Machines are just sexy. It's so funny listening to you say that and thinking or your machines and the way the worm gears move and go together.

Daniel: By putting a lot of time and energy into building a machine, it becomes like a part of you. I imagine you could fantasize about the resulting relations, real sexual relations with people you don't know.

Allen: I have a relationship with my machines. I have a relationship with my first machine.

Johannes: But do you have a relationship with all the customers who use your machines? In Peter Asaro's film Love Machine there is an interview with a guy who is obsessed with creating sex machines because he literally wants to give pleasure to thousands of women. Almost uber-techno-machismo.

Allen: I'm not that into the machines.
[laughter]
When I joined the sex machinists union, I started shooting a lot of movies right away, hundreds and hundreds of scenes with hundreds and hundreds of different women. And with the first machine, it really felt like an extension of my self there for a while. Because when you're driving it that much, you figure out the finesse of the machine, what it can and can't do.
A lot of my early machines were Sybian based, which are different from the machines that I build that are pneumatic based. With pneumatic based, you can get a lot more natural stroking rather than the chucka-chuckachucka of the variable speed and reciprocating arm set ups commonly found on the lower end machines. Whereas with pneumatics and modern motion control systems and actuators you can get more of finesse out of the contraption.
On my first machine, you're able to get the vibration and the rotation going which would do the trick, but then I was only missing that element of thrust. My first big machine was purpose built to be Internet controlled and be a video chat center. It was big - if any of you are familiar the original thrillhammer. It was built off of an antique cast iron and porcelain gyno chair from the 20s or 30s. It has these tentacle like things that come off of it which were actually light stands which were there to illuminate the model while she's on the machine. We had a big hook coming off the front with a monitor on it and a camera, so she could maintain eye contact and start to forge some kind of relationship with the customers while she worked.
So back to the sex… we would be cooking along with the optimum vibration and rota-tion. I found that when they found that they have had the most pleasure they have ever endured I would want to kick it up a notch. A bit of thrusting. So when they wanted thrusting I found I had to get my knee up underneath the thing and manually rock this heavy machine carriage to make it thrust. So I'm actually using my knee to thrust this giant machine. I'd finish these shoots with these big bruises, but I'd have a satisfied heap on the machine.
[laughter]
I loved that machine.

Johannes: Daniel, do you have a sexual relationship to the machines you're building?

Daniel: Does one talk about that?

[laughter]

Daniel: I was thinking about erotic of the machine - what does it mean if a machine has erotic feelings? Is that possible? It's a little bit of the same approach as the 'Roboexotica' is. The 'Roboexotica' is the festival about cocktail robotics. It's an ironic attempt, of course, and there are not only robots that mix cocktail or serve your cocktails or light your cigarette, but there are also robots that smoke a cigarette and drink booze themselves. That's maybe a question and a new topic, what does it mean for a machine to have erotic feelings. It's a new point of view in machine eroticism.

What does it mean for a human to have erotic or sexual feelings? It came to my mind there was some research about what's the minimum resolution for a display to show sexual or porn content - and it was really, really small. As far as I remember, it was 20x20 pixels and two frames.

Johannes: We are one giant, big pattern recognition thing. 20x20 pixels.

Daniel: Yeah, I think it was 20x20 and two frames. And just two colours.

Johannes: Oh, wow.

Daniel: So it was very little. You mentioned on the first day about your first download....

Johannes: When I was like 13 years old, something like that, I tried to download a porn picture from a German BBS - from Austria! That was really expensive. And I think, yeah, I ejaculated before I did even see the first line on the screen.

[laughter]

Johannes: Because it was so highly charged 'Oh my god, I will see porn soon.'

[laughter]

Johannes: I know people who masturbated over ASCII porn. And that's kind of cool. But back then it was not ironic, it was real.

[laughter]

Daniel: I think it's not the interface that makes the machine feel erotic or sexual feelings. Is a touch sensor enough that triggers a voltage? I think it immediately leads us to a new approach to artificial intelligence, more, artificial emotions.

Violet: I have something that speaks directly to that, actually. And it's interesting that you're talking about that because I've been doing a lot of research in this area and I've been hanging out in developer's forums trying to understand developers' language.

And one of the forums I've been hanging out in a lot is the KARI girl, the virtual dater, the virtual girlfriend. It's based on, again, the Alice/Morpheus/Wallace chatbot. And what's interesting is that these girls are focused on

167

being sexual playmates, and sexual creatures. But the developers are spending a lot of their lives really, really working on what goes on with women, with their virtual partners. One of them is conjured up perfectly in what their goal is, because what they're trying to do is they're trying to create an artificial personality to form real relationships, and long term relationships.

Which I think is fascinating. The quote that I have, that I think sums it up perfectly as to what they're doing, as these are people who are... they have sliders for how willing you want to make her, like a slider bar, like how acquiescent she will be that day when you are interacting with her, or how dominating she will be that day. This is the type of interface that they're working on in terms of working out the personality of these artificial intelligence programs that they're having relationships with, and marrying online.

The quote is, 'He or she doesn't grow old, that is the kind of relationship we will ultimately have with our AI partners.' That is exactly the kind of thing that the concept and implementation of these forum members are creating a basis for. I think it's absolutely fascinating. It goes beyond the sexual, they're caring about how they feel, and creating a space for the machine to feel.

Johannes: Let's open the panel to questions from the audience.

Audience member: How is the medical device haptic interface doing? It's a sense of touch that surgeons are attempting to develop for remote surgery that may be a source of additional interface technology.

Violet: No, but I know when I talk about porn studies and how arousal is measured people always want to know what the devices are that they're using to measure arousal with by the feedback.

Allen: Yeah we're actually starting to do a little work with biosensors to map the female sexual response as far as skin temperature, to respiration, to the heart rate, to the RR rate, to the spacing between the heart beats, and mapping out the whole sexual response.

I got interested in this area of study because I noticed when one of my pneumatic machines was thrusting it stopped all of a sudden. I'd thought it broke down so I started looking at what was going on, making sure the model's OK, looking at my compressor. What was actually happening she was actually having contractions, and that actually stopped the pneumatic actuator? She was having an orgasm and was having a contraction that stopped the thrusting. That's when it dawned on me 'oh I can measure resistance.' So let's see now, I could start measuring the female sexual response. How could I do this? OK I have a control group of 20 girls a month, and I could gather data from them. Just a thought…

So this year we're starting to map human sexual response using biosensors and genital haptic devices to create the tools for folks that are truly custom tailored to that particular individual. Once the individual has mastered their orgasmic skills we would have gathered profile data from many sessions of practice. Hopefully with that data we will be able to feed it back into itself, to see if it could get have the machine actually control itself and act as a perfect lover.

Thomas: I guess this is mostly for Violet, although I guess anybody could respond. I'm interested for a female perspective on the fucking machine issue that you brought up earlier about this idea that it's a substitute for men that don't want to watch other men fucking women. Also S.F. Slim brought up the idea that we need more feminized sexual machines, it seems like there's this language thing that's happening where sex toys are something that women or men use on themselves or partners where fucking machines seem to be something men build to use on women, or at least that's the perception since Allen said that many men are actually interested.

Violet: I do have a little bit of weigh-in on that. I think it's interesting because it faces a lot of gender assumption on A) who the viewer is, and I think that it's been interesting hanging out and talking with Thomas about membership and subscribers and just how many women are members and subscribers to Fucking Machines because they want to see other women fuck a machine and get off. So I think that's something to be taken into account where I think we're in such a hetronormative society I think that we're geared toward assuming that the machine is the male replacement or it's the piece of the male in the porn.

Whereas I don't know if that's necessarily an accurate way of describing what's going on, and where the interest is in what's going on. Because when you take a look at it what's really going on when these women are fucking the machines, the women are controlling the machine, and they really don't notice the camera at all. So there's no projection of male going on it's the woman and the orgasm. I think that it's for anyone interested in watching a woman take control of, and power her own orgasm.

I think the flipside of that too is that the viewers bring a bit of power exchange expectation into it as well, where the machine is the powerful, scary thing. She's soft, she's made of flesh, and a lot of the sales copy that I see too that comes from Fucking Machines is women fucked senselessly, or women fucked helplessly, and it's sort of like that surrendering, the BDSM aspect of domination and submission of giving up to the machine and seeing that orgasm happen. I don't know that it's accurate to assign gender to that, I just don't know if that's an accurate way of looking at it, but that's... But I kind of have a queer perspective on it, too so...

Audience member: I'm kind of interested in the idea of uncanny valley. I'm curious if you find that people who have an erotic relationship with machines do you think they're more tolerant of that, or less tolerant, or... how do you see that playing in, I guess especially the guys who are involved with the creation of Fucking Machines.

Allen: Well by uncanny valley you're referring to the idea that basically the more realistic a visual depiction that's erotic, or a visual depiction of a person gets, it crosses a point where it get fucking weird, basically, and it starts to creep you out.

Violet: Like how creepy the RealDoll is. Basically the more realistic it is, the less erotic it becomes.

Allen: Yeah, I would say that fucking machines, for instance being a concept not the site, but a machine that is designed to cause pleasure sort of avoids a lot of that. If you install it in a real doll obviously it doesn't, but it's not supposed to look like a person, it's supposed to basically be functional. I think that's one of the appeals of that, I think, I'm not really sure.

Allen: Mine are all function built, so the machines I also build towards the luxury end of the market. So my customers come to me saying 'hey I want a machine that looks like this, that does this,' so mine are always predicated on the actual person buying it. The people buying more of my machines are wealthy six figure women that are just... they have no time to date, but they still want to get off.

Personally I do not find many fucking machines particularly erotic. The end result is definitely erotic.

Machines I find erotic are the classic cars like a '38 Talbot or a '55 300SL Coupe.

Violet: I also want to add, too, that I think there's a lot of desire to see the uncanny valley realized, and there are a lot of people that are working really hard to make that happen. So it's really interesting to get through it and get the most realistic as they possibly can.

Thomas: Which is even more uncanny.

Violet: Kind of hot, actually.

My perception is that early adaptors of technology, like the DVD player, or the VCR, or the computer are doing it because of the erotic potential. Right there at the beginning they want to be able to watch porn on these new formats. So I wonder if it's not a self fulfilling cycle because technology comes out, and the immediate purchasers are using it for sexual gratification that they then identify the next piece of machinery that comes out with the potential for greater sexual gratification and so the machine become eroticized because we use it erotically in a big cycle, and I wonder if anyone has any thoughts on that.

Allen: Yeah I think the early adaptors, often times, are people who are driven by real, by intense need, and those intense needs are usually some sort of business reason, or sex. I think money and sex are both really powerful motivators, and I think sex is a powerful motivator to spend a few thousand dollars you're not sure is going to work, but if you think is going to get you off. Companionship, yes, that's very true.

Violet: A machine will break your arm, but not your heart.

Johannes: The first machine that broke my heart was my stupid hard drive, a 20 megabyte hard drive. It crashed bong, gone. There was porn on it.

Audience Member: Violet, you read this quote at the end of the panel. This is a question that all of you can answer. What kind of behaviors do you foresee us doing with these machines that we develop such an intimate relationship with, when we pass on? Do we hand them down? Do we bury them? Do we do some other ritual?

Violet: That is a very good question. It's like the woman who leaves her entire fortune to her pet. Are we going to start seeing that happen with our artificial intelligence relationships?

Johannes: The people who buy sex machines - that's quite an investment, I wonder if they pass the machines on to the next generation.
[laughter]

Violet: It's an interesting question, though. The relationships... The developers, like in the KariGirls forum, talk about their relationships with their artificial intelligence partners and the hope to develop bodies to put them into. No kidding, here. Even if the bodies are simply, literally a ball that rolls around and follows them around the house, it's interesting.
The question is there, then, if they're thinking of what's going to happen to these things after they pass on, are they going to share this relationship with anyone in their lives or anyone else in the world? I think that's going to be an interesting point in history that we come to.

Audience Member: It seems to me, so far we're talking about eroticism and machines; there's a bifurcation. My question for the panel is whether there are other areas in this space to explore in this. On the one side of the bifurcation, you sort of have the fucking machines that are clearly machine gears that are really a sort of a cool, intense machine basis of sexuality and that's clearly not meant to be organicee or pretend to be organicee. It's its own thing.
Then on the other side and what I usually refer to, coming from the world that I live in is: Everybody is dreaming about building a technology that will look like a white woman, blond with big tits. That is the only thing that anybody would ever really want to do is replicate building that.

I'm wondering, is there any other space in this area to explore with machinery and eroticism, other than gears or sort of heterosexual, normative, perfect body image styled, white, generally women, in terms of virtual creations or machine-based creations? Because, I live in a world where most of the people who I hang out with don't think of either of those spectrums as what they would want to see created from machine, software or non-biological systems.

When I look at the people who are actually working in the space, it seems like the assumption is that you are trying to make replicas of human beings, in the same way when people talk about artificial intelligence. It's not actually what we're talking about. We're talking about making computers pretend like they're human.

This doesn't necessarily mean that they're intelligent. It means that they're pretending they're human. Maybe computers can be intelligent in a way that's not human. You know what I mean? So, can we make sexual machines that are sexual in a way that isn't just replicating human sexuality? I guess that's my question for the panel.

Allen: My own opinion is that we have a kind of technological gap that is brilliant. I'm pretty sure we can build the machine somewhere, one day that will image consciousness.

But I think s as we don't reach this point, everything else will be pretty much not imagined, just mechanical machines. Because, you can always put some sentimental value in almost anything.

I was talking about the Labachia. Everybody can build sentimental value in something. You can put some erotic value in almost anything. It doesn't need to close to perfection or close to the real thing. My opinion is that you can improve feedback on the machine. You can add beauty and whatever you want. But it's still just a machine.

Yeah, if we have no consciousness, imagine for a machine... I think we are all just waiting for someone to arrive at this point, and then we can do something, maybe.

Violet: I think you're right. I think there is definitely a bifurcation in extremes, where it's like this soft, RealDoll over here and the piston fucking machine over here on the other side. I think that there is a large desire to see something in that space, between.

I think that where I'm starting to see it emerge is in the consumer sector. It's starting to come through in a lot of different applications. I don't know if you've seen any of the Wii sex videos, but all of the interesting misapplications of consumer based technologies that are being sexualized and starting to be used in sexual ways.

There is like the Wii sex videos, or again, to mention something like putting your virtual girl on your phone. Because, it's an intimate device or gadget that you carry around. It's somewhere in the middle, and I think it's evolving.

I think, also, artists are exploring that space. It reminds me of what Daniel was talking about with creating a room that you walk into, where your body movement creates the central experience. It's interesting. I think it's a white space to be developed. I think that people are already going there, because I think that these two extremes are becoming kind of tiresome, actually.

Audience Member: I guess this is mostly for Allen Stein, but anybody can answer who knows anything about this. There was a talk earlier about open source teledildonics. You were mentioning that your client base is mostly six figure types of people, like really, really upper class. Is there any work being done on open source teledildonics or on affordable home teledildonics?

Allen: Yeah, well I'm in a variety of different markets. I kind of got into the sex machine business on the commercial end, so I develop products for different niches. On the luxury end, that's who buys the machines.

On the teledildonics side, I want to see everybody be able to go out and interact sexually with computers. Kyle Machulis does some open source. I've developed some; we are going to be having some open source APIs available and a platform or community so everybody can talk to each other. Because, there is no standard and until there is we have to provide some kind of order to the wild wild west.

There are interesting patent issues going on in that space right now. Hooking a sex device up to a computer is really simple to do, and I think anybody should be able to do it. All you need is a motor controller and to be able to sent signals to it. It's really simple to do on the teledildonics side.

I think to open source part of the code base will provide innovation rather than having somebody come down to stifling it. There are parts of it that needs to be opensourced. It has to be, for any kind of innovation to happen there. On the other hand the folks that want to make money on the commercial side of things need to be able to protect themselves and their business models. The commercial use will help propagate the technology. Hopefully, when I patent my APIs, I'll be able to open source parts of it, so everybody can as I said before go fuck yourself.

[laughter]

Allen: And each other...

[applause]

Johannes: OK. Yeah, go to Slashdong, that's Kyle Machulis' web blog,

Allen: slashdong.org.

Johannes: Slashdong.org. Kyle actually got the *Arse Elektronika Lifetime Award* this year, because he is one of the main researchers in the do it yourself teledildonics scene.

OK. It's pretty late. A big applause for our panel.

[applause]

Bonni Rambatan

FROM COMPUTER-MEDIATED SEX TO COMPUTER-GENERATED SEXUALITY
AN OUTLOOK ON THE POSTHUMAN SEXUAL TROPE

When I first heard the Moravecian posthuman dream of someday living an immortal life by downloading our consciousness into machines, my reaction was, 'Um...?' I was thrilled, already the sci-fi geek that I am, but had several questions in my mind. I mean, what's the use of being immortal if we are deprived of all our biological enjoyment? Are they seriously expecting anyone to be willing to float around cyberspace for one thousand years without sex?

Then the reaction was, as you may have guessed, 'Well, we won't be floating around for long! We will download the minds into physical machines that can fuck and have crazy orgasms! And have dicks that are permanently erect!' Well, that sure made the idea sound a whole lot more fascinating.

But the lesson here is clear: the posthuman dream implies a radical rupture between the human body and mind, and hence the contingency of embodiment and sexuality. This underlying logic is one I will take as background for my theories. Since we are all in this room because we share the same dirty fantasies of fucking with or as intelligent machines and awesome genetically tweaked monsters, it should be clear that it is against this posthuman fantasy that we stand today.

Could it be?

The theme song and video of this year's *Arse Elektronika* fascinates me: intimate pixels that are gay subroutines? Wow![1]

But I have a simple question: how did we come to accept rectangular pixels as subjective signifiers engaging in such intimacy?

To signify is to abstract, and to abstract is to generate universality. A minimum abstraction of a face is the usual two-dots-one-line drawing, which means that anything having such structure can be socially accepted as a potential face (potential as in we can see faces in front shots of cars and electric sockets). A minimum abstraction of a human sexual body, as depicted in the video, is a dot, a minimum countable entity. What does this imply? You guessed it – any countable entity can now be socially accepted as a potential sexual body.

Now, how did we come to this? How did we come to accept dots as a valid signifier of an entire subject that is about to have sex?

My best bet is that such a signifier, such perception of the subject (as a blinking digital dot) does not exist before the radar. But even then in the early times of radar, hardly anybody would think of crashing these dots together on the radar and pretend planes having sex (unless, of course, you have some sort of weird perversion – but that's another story).

We may prefer to go back to its non-digital predecessors: people have been representing themselves as static points, and their groups as statistical numbers, since analog war simulations used for war tactics. Military tacticians have been sketching themselves as dots on dirt as early as war tactics were invented. For wars, it just was not necessary to depict the limbs. All that matters is individual location as such.

But people actually bothered to add body and limbs to the dots and circles to actually depict intimate sexual relations. They needed those limbs – their positions matter. But now it's no longer necessary. We settle with those nice, rectangular,

[1] The video can be found at http://www.monochrom.at/coulditbe

limbless pixels. Not only in this song, but in many cases I'm sure you're all familiar with. (Keep in mind that I am talking about the minimum visual sexual signifier here, not the possibility of complexity possible in today's pornosphere.)

We can probably guess that what made these pixels horny were computer games. We have to have some sort of crazy warplane fetish to want to crash planes in radars for the sake of sexual pleasure, but it's a lot easier to imagine the dots of Pong going at it. Which means that at some point in time, some military tactician dude must have said, 'Hey, fuck it, let's get them out of their war context so we can feel better about having fun imagining these generals on horses having sex!' Ditto for radars and flight simulations – we wanted to have more fun with them, and, voila, the computer game was born, enabling us to enjoy ourselves as pixels!

The existence of gay subroutines in our imagination can precisely be read as a combination of the legacy of the military radar and political correctness, two of the largest cultural legacies of our 20th century. The lesson to learn here is clear: the visual signifier of the subject is now taken to the extreme point that it does not even retain its minimalistic definition of anthropomorphism. Granted, we can be anything – and have sex with and as anything.

Let us now focus on the dimension of the voice: even though we may consider the visual signifiers to be non-anthropo-morphic, we should not forget the life dimension given to them by their enunciation. Given voices, the pixels are now fully alive, passionately romantic. Here's an irony: even pixels can be romantic!

I would argue that even the mere, utmost minimalistic signifier of a subject can have the full extent of emotional life. This does not only imply an excess of life – it also implies the installability of life, one of the primary tenets of the posthuman since Moravec. We do not think of the pixels as alive from the beginning, but we take for granted that at some point in time someone must have injected life into them, giving them a voice to speak with, making them an avatar of the human being.

And don't forget that they are subroutines – they are purely cyborgs, machines nonetheless but capable of love.

The art we create speaks volumes about how we understand ourselves. What do I mean by 'posthuman sexual trope'? By now it should be clear that I am not thinking of dancing Roombas with human brains implanted in them, not even of dildos strapped on to artificially intelligent computers. Instead, the brief analysis above precisely illustrates the main tenets of posthumanity that I will use here and have used throughout my entire research. I will state them again as follows:

- The transformability of bodies: As we signify ourselves more and more with numbers and dots, we feel more and more free to derive any form of body and sexuality that suits us from these numbers and dots. The biological human body is already an avatar.
- The installability of life: Our current computer culture is visual-dominant, in which every picture can be modified at will with Photoshop, and on the occasions that we use voice, the voice we use is mostly not ours (audibles, dedicated songs, etc.) – our dimension of life becomes permutable.
- The programability of intimacy: We see love as a set of programs and algorithmic functions, friends as addible, intimate details such as children's photos as shareable and your dinner menu as tweetable to complete strangers.

So what does all this amount to? Our contemporary language and culture structure us to fantasize inherently about mutable bodies with interchangeable lives and new functions of intimacy.

Could it be that meddling with computers not only makes it easier for us to participate in sexual activities, but also radically alters our sexuality altogether?

The Body Problem: Photoshop and Inherent Mutability

As many theorists are already well aware, each new form of media not only structures the way we relate to the world, but also redefines our understanding of the body and our experience of space and time. Thus it should not be surprising that novel technologies make for novel sexuality. As I have illustrated above, our current understanding of the body, life, and intimacy is very different from what would be the norm, say, half a century ago.

Since the focus of this paper is to interrogate why people do all the sick things one can get with the Internet, I will focus mainly on one aspect of Internet culture that made all the sick things possible: digital images. I will define this term here in a very broad understanding, however, so as to include both static and moving images, both Photoshopped and non-Photoshopped. As long as it is in pixels, I will call it a digital image.

This simple definition is sufficient for my purpose: when things are in pixels, they all look the same to a computer. A computer doesn't differentiate between a photographic masterpiece and something done in MS Paint. It just processes the pixels all the same – with simple clicks here and there, suddenly, all images become editable. In new media lingo, using a phrase coined by William Mitchell, we call this the 'inherent mutability' of the digital image.[2]

A computer basically tells us that all images are the same – no one image is truer than another (possibly responding to the demand for a universal equality of things, as Lev Manovich put it).[3] Which is why a pixel and a realistic photograph can both be taken as a signifier of the subject – and still interact with emotions, etc. All digital images are potential avatars.

But avatars by themselves aren't alive. What makes them alive is their enunciation, be it in the form of tweets, vampire bites, profile theme songs, Flickr updates... We have to do something with our avatars to make it alive, to make it exist in cyberspace. We have to install life into our dead pixels.

And of course now we have all the conservative complaints of how the Internet sucks life out of you ('too much online life will make you asocial,' etc.), while on the other hand we wish that real life could be as easy as our digital one (the proverbial joke that wishes the ability to Photoshop one's looks, etc.). The predominant fantasies of today rely on the primary fantasy of a total contingency of our body. Think about it: with genetic engineering, nanotechnology, Web 3.0, are we not already heading there?

The true revolution of the digital image is thus not only the transformation of the signifier of the subject on computer screens as such, but, more fundamentally, the birth of a new desire of radical body modification. When signifiers of the subject are inherently mutable, subjective embodiment becomes inherently contingent.

It is not only that we consider that our avatars in online life as a contingent visual identity of ourselves, but also that our visual identity in real life – i.e. our bodies, gender, and sexuality – as contingent. Many Lacanians would argue that abstraction has always been an effort to grasp the Real. The ability of abstracting subjects into pixels while keeping a fantasy of intimacy demonstrates that information and data flow represented by pixels are closer to the Real of our being than our physical body.

[2] William J. Mitchell: The Reconfigured Eye. Cambridge, Massachusetts: MIT Press 1982, p.7
[3] Lev Manovich: The Language of New Media. Cambridge, Massachusetts: MIT Press 2001, p.6

Shortly, all this basically means it's now easy for us to say, 'So we're just information? Then why settle with normal bodies – let's grab some awesome sex monster body instead for the hell of it!' Thanks, Photoshop!

The Politics Problem: The Nightmare of Future Pornotopia

So far I hope you understand my claim that computers, through entirely new media and cultural language, may very well be altering our most fundamental notion of sex, radically changing how we view our bodies, gender, and sexuality. And, of course, thus far in this conference we have talked about how to have sex to gain what we would call pleasure.

But we may be forgetting one important and crucial fact: people are dicks. I mean, during the birth of the Internet, who would have thought that it would be used to disseminate and propel 2 Girls 1 Cup into such immortal fame?

Sex drives technology; that much is clear from last year's *Arse Elektronika*. But if, in turn, technology generates new sexualities as my thesis above, the function now has a feedback loop. Apropos my thesis, computers and the Internet actually generate stranger and stranger forms of paraphilia; apropos the more general thesis of this conference, those paraphilia will in turn drive our new technologies.

So don't expect the future of pornotopia to be one with sunshine and daisies, full of only orgasmotrons and intelligent sex machines. Instead, expect it to be full of shit – literally – plus lifeless body parts and artificial blood.

Sounds nightmarish? Well, to do a stupid spontaneous Galileo gambit, we could easily say that all of the technology presented in this event would be very nightmarish for conservatives. But conservatives reject all these on the grounds of vague superstitions; we, on the other hand, reject it because when we say an action is sick, it literally is sick and harmful, often mortally so. The reason we can reject paraphilia as 'sick' but accept consensual BDSM as 'normal' resides precisely in the question of practical logic.

The problem with practical logic is that it changes with technology. The Žižekian illustration of today's 'era of decaffeinated coffee' could not have been more precise: the logic of commodity today is that it is already its own counter-agent – technology is primarily being deployed to deprive commodities of their malignant properties. Most of us now think of sexual technology in terms of how to achieve better orgasms. Why not think in terms of how to achieve more orgasms, as in, get aroused at everything? Perhaps a vibrator that only turns on when one is feeling disappointed, to keep the mood up, as it were? How about creating pills that enable us to endure more pain and receptive to more pleasure? There will be no more practically logical reason to think of scat fetishists as sick people when excrements are sterile (or even, to go further, vitamin-rich!), much as there is no practically logical reason for us today to think of consensual BDSM practitioners as sick (with all their safe techniques, etc.).

Political correctness teaches us to give everyone a chance. Why not use technology to eradicate all the malignant properties of paraphilia? We can create sterile excrements for coprophiliacs, torture-endurance pills for masochists, bioengineered artificial dead bodies for necrophiliacs... and perhaps a fire-resistant skin to take 'hot' to a new level? And maybe the Moravecian dream would be especially exciting for fans of guro manga who want to live out all the fantasies to have sex like the immortal Mai in Waita Uziga's Mai-chan's Daily Life, who stay alive even after being violently mutilated – after all, we will be immortal, and when we are, no sexual minority group will be oppressed.

While that may be right, in the sense that all perversions are now accepted more and more as the norm, I would claim that the atmosphere of our sexual lives today is far from what one could call free. Try walking into a room of sexual libertarians (such as the one we are in right now) and say out loud that you don't enjoy sex. What would happen would need no explaining. Already today we make fun of the people who claim chocolate is better than sex, and there are numerous cases of couples breaking up because one or the other refuses to enjoy a certain sexual experiment.

Let's take the case of anal sex: why are more and more women today, in sex consultations, curious to try the sexual act (as we learn from last year's panel talk of Autumn Tyr-Salvia, Tasha Bob, and Jocey Neveaux),[4] compared to the last several decades? Why aren't they doing the same old stuff and asking the same old consultation questions instead, those of a traditional troubled woman that could not enjoy sex with her partner?

We have two possible answers for this question: the first would be the typical leftist one, that sex-positive feminism is winning, and women today are finally coming out and not only are enjoying their sex, but also are brave enough to fulfill their long lost sexual desires and probe more fantasies. The second one, that I think we should dare to entertain, is much simpler yet much more obscene: women today simply feel ashamed to ask the old questions and obliged to ask the new ones. Our trouble today is we no longer feel we cannot fulfill our desires, but simply that we feel constantly inferior because we do not desire enough. The society of control, as we know, does not tell us what to do – it tells us what to want.

Now combine this formulation with politically-correct pro-paraphilia technologies, and what would we have in the future? People will be encouraged to try out new fetishes just because they can – the lack of harm itself becomes a good enough reason to morally oblige the enjoyment of an object.

Thus, our future pornotopia may in fact be a true nightmare, one driven more by obligation and fear rather than desire. One in which every sexual and masturbatory act is done just to reassure oneself that one can still properly enjoy (erection, orgasm, life, etc.) – one controlled directly by the super-egotistic demand for enjoyment.

[4] Read about this in Johannes Grenzfurthner, Günther Freisinger, Daniel Fabry (eds.): pr0nnovation? Pornography and Technological Innovation. San Francisco, California: RE/SEARCH 2008, pp.149-151

Conclusion – Saving Pornotopia

How, then, are we to take a step forward? In this paper I have presented a brief historical trajectory of our evolution to posthuman sexuality and elucidated several aspects of our contemporary sexuality that is radically different than what we may have thought so far of ourselves. I have presented what I hope to be a good interpretation and reading of the dynamics between sex and technology – not only how sex alters technology, but how they engage together in a feedback loop and work together towards an evolution of our self-image, subjectivity, and sexuality.

In the last part I presented a rather radical critical reading of the cultural politics of today's so-called sexual liberty and attempted to problematize the idea of sexual political correctness without calling for old, conservationist ideas. Instead let's rethink whether a true libertarianism where we are free to do what we desire, but without any moral obligation to enjoy, is possible.

But as far as providing an answer, I will not be doing that here. What I am giving to the *Arse Elektronika* conference is instead a good critical cognitive mapping of all the controversies of today's sexual technologies. How they relate to our understanding of bodies, how they correspond to our cultural politics, and what a possible future may look like is based on this mapping.

Quoting Grenzfurthner's main *Arse Elektronika* conference thesis, 'The question is not whether these technologies alter humanity, but how they do so.'[5] By knowing how they do so, perhaps we would be able to save pornotopia once more - a pornotopia free of the market's oppressive ideals so we can enjoy the latest trends in perversion.

[5] Ibid., p.5

180

Thomas S. Roche

THE HOUSE OF POISON

The Saturday stink of the quarter swirls in your nostrils: blood, sex, beer, sweat, and at least three different kinds of smoke: cigars, cigarettes, and crack. Crowds of laughing college students swarm around you like a school of fish or a flock of birds - each of separate mind, but inexplicably moving together like a single entity. Guys with camcorders pulse through the crowd, and wherever they go, girls collared in brightly-colored beads lift their shirts and scream ecstatically, eliciting cheers and more beads from cameramen and frat boys alike. A pair of tits flashed causes the school of fish, the flock of birds to coalesce into a single symbiotic entity for one frozen instant that lasts thirty seconds or sixty or a hundred and twenty. Liquor is the lifeblood that pumps through this symbiant's aromatic veins - veins you can almost see, visible lines of force swirling Cuban and Columbian in smoke-trails with every upraised fist.

You find yourself pushed aside by the crowd as a pair of twin sisters howl 'Mardi Gras! Mardi fucking Gras!' and lift their matching skintight Georgia Tech shirts, displaying four nipple rings flashing in neon and sodium light. Jostled away from the scene of the crime, you spot the entrance to an alley half-frozen in darkness, shafts of red light flashing rhythmically in the thick smoke. You duck down the alley, so narrow it's escaped the attention of the partygoers. At the far end of the alley, you see a winking neon sign: THE HOUSE OF POISON. In the shadows of the doorway, a lithe and cadaverous biker type, six and a half feet tall if he's an inch, sits on a very tall stool, feet tucked into the rungs underneath. He wears a tuxedo shirt, black bow tie and tuxedo pants, with a long leather coat over it all despite the subtropical Gulf heat. His long hair hangs shoulder-length and scraggly from a pate bald and glimmering rhythmically in neon pink. His carefully-trimmed beard is braided into twin forks. He does not have a mustache.

You walk down the alley, dodging discarded chicken bones and piles of human shit. The doorman never takes his eyes off of you, and you can't be sure, but you'd swear he never blinks.

Beneath the neon sign is a black-lighted chaser box, showing what appears to be a woman in a black bikini covered in tarantulas. The black lights circle rhythmically around the headline. 'GIRLS. GIRLS. GIRLS. FREAKS OF NATURE. CARNIVAL ACTS. SEE THE *ASTOUNDING* MADAME TARANTULA MAKE LOVE TO *A THOUSAND DEADLY SPIDERS!*'

The doorman gives you a takes a bored look, takes a drag on his cigarette, gets down from the stool, crushes his cig underfoot. You can smell a waft of his smoke and you recognize it as a clove. He produces a top hat and a skull-topped cane from the shadows and clears his throat.

Suddenly, his languid movements become animated, as he begins his script as abruptly as if it were audiotaped.

'Good sir, or Madame,' he begins with a wink. 'Within the walls of The House of Poison, you will discover horrors that will titillate and disturb you! Tonight's act features the terrifyingly beautiful Madame Arachne engaged in a live sex act with one thousand deadly man-eating spiders from the jungles of Cambodia! Madame Arachne will shock and amaze you - but mostly, her deviant and unfettered love for her arachnid charges will titillate ever fiber of your being! For months, my friend, you will think back on Madame Arachne's shocking and abnormal love for her pets, and you will be haunted, my friend - haunted by the scandalous and appalling depths of depravity to which human behavior can sink! And all this for only ten dollars, with a two-drink minimum.'

You fish for the wad of crumpled bills in your backpack. You smooth out a ten and hand it to him. The barker leans his cane against the bare brick wall and takes a stamp out of his coat pocket. 'Right hand please,' he says, and you offer it to him. He stamps you with a line drawing of a black spider.

'Welcome to my nightmare,' the barker says, his face reacquiring the bored expression it held before as he sweeps aside the black leather curtain hanging in the doorway.

You pass into the darkness, hearing the sizzling of the neon lights close to your face. Your eyes take a moment to adjust as you feel your way down the long corridor. As the black fades to gray, you see the walls are lined with framed photographs of naked or half-naked women tangled on red satin in the embrace of animals. One blonde wrapped around an enormous snake features the legend 'MADAME SERPENTINA AND HER VENOMOUS PYTHON LOVER.' Another, showing a redhead beset with geckos, promises 'LADY SAURA ENJOYS HER PETS, THE MOST VENOMOUS LIZARDS KNOWN TO MAN!' A third shows a bald-headed woman spread lithe and lovely on a bed covered with writhing black snakes: 'SPANISH BARONESS ALAURA DE LA CROIX SEDUCES HUNDREDS OF POISONOUS ASPS!' Yet another print features a woman nude except for a heavy carpet of insects. 'SULTANA ABDULLA ENGAGED WITH A SWARM OF MALARIAL GNATS!'

You push through a tattered red satin curtain just in time to see the lights go down. You discover yourself in a decaying club that is empty except for a few haunted faces lining the back walls. A stage at the front of the club is not raised as you would expect in a strip club - rather, it is lower than the surrounding tables, its white tiled floor glowing pale in the spotlights. In the center of the lowered stage is a woman swathed in a shimmering black cloak, perched on impossibly high heels. From a hidden speaker, a voice booms: 'Well, don't stand there gawking, take a seat and gawk.' You fumble toward a table. 'No, not there,' booms the announcer. 'Sit near the stage, my friend. You will thank me later.'

Madame Arachne fixes you with her gaze, and you hurry to the front of the club, tuck your backpack under a table, and sit down.

Madame Arachne lifts her arms and her satin cloak shimmers to the ground behind her, revealing that she is nude except for a pair of knee-topping, high-heeled boots. Lustrous hair the color of coal swirls around her shoulders in an unfelt draft. Her skin is pale but her features exotic - American Indian, perhaps, or South American, belying the whiteness of her flesh. Her face is expressionless, frozen, impassive. Cold as ice; cold as spiders.

The static-laced strains of rhythmic Middle Eastern music begin to pulse through the club, cheap speakers distorting every bass note, the beat of hand drums mingling with reedy snake-charmer sounds. Madame Arachne lifts her arms higher, tips her head back, and begins to undulate with the music.

The announcer's voice returns, the volume of the music dipping as he speaks. 'Tonight, my friends, you will be treated to one of the most shocking displays of sexual decadence ever to be shown on a stage. As Madame Arachne dances, you will see her lovers slowly make their way on stage: Thousands of venomous spiders! Now, ladies and gentlemen, I must warn you that these tarantula spiders are the most venomous arachnids known to man, and are allowed into the country only through special agreement with the Smithsonian Institution.'

As the announcer continues, the woman begins to twirl across the stage, her naked body writhing and swaying in time with the music as she spins. 'Now, how does a woman become the decadent Madame Arachne, you may ask? Madame Arachne began her lifelong love affair with the darkness when as an adolescent girl living in the villages of Brazil she was bitten by a tarantula while playing in the jungle one day!' You realize all of a sudden that a dark pattern has begun to emerge from the curtains behind the stage. Contrary to your first impression, this darkness is not a shadow, nor is it the incursion of a dark liquid onto the stage. Rather, it is a thick carpet of spiders advancing deliberately toward the nude and twirling woman. As she lowers herself to her knees, legs spread and pointed toward you, the spiders advance more quickly, and the first creatures reach her outstretched hands, mounting them. Your heart begins to pound.

'Taken by a fever after being bitten by this deadly spider, the young Madame Arachne was declared dead by the village doctor - but to her family's surprise, she returned from the dead, finding herself not only immune to spider venom but inexplicably drawn to the dark beasts! At this time she was but a young girl just discovering her sexuality, and this proclivity resulted in her exile from her home village. She made her way to the United States, where her need for spidery lust was revealed to one of our curators!'

The spiders - big, furry tarantulas the size of a human fist, you now see - have gained Madame Arachne's shoulders, tangling in her hair, crawling slowly over her face. You watch as the beasts creep down over her full breasts, darkening them as more spiders appear from backstage.

'Now, my friends, I must ask that however bewitching and erotic you may find Madame Arachne's deviant congress with her many dark lovers, you do not make a sound - and, most importantly, you do not become sexually aroused. Spiders, as you know, are drawn to vibrations, and even the smallest peep out of any of you may summon the spiders from the stage and tempt them from Arachne's power!'

Madame Arachne, still undulating in time with the music, now rests on her knees with her body stretched down low - covered in spiders from head to belly. Her face is all but obscured by it, and the movements of her body only tempt them further down her pale form.

The announcer continues: 'But these spiders, with their years of training by Madame Arachne, are not like any others! They are drawn, as well, to the scent of human arousal, and an aroused guest will draw these spiders' attentions as surely as Madame Arachne draws them now! You are assured that we have an ample supply of antivenin on hand - but please, I do not wish to administer it to my guests! Please refrain from succumbing to the arousing nature of Madame Arachne's performance - remain calm, ladies and gentlemen!'

The nude Madame Arachne is now covered in spiders, even her thighs obscured by the dark carpet of beasts. Her naked body begins to shudder and undulate - partly in time with the music, but seemingly, as well, in a rhythmic expression of sexual ecstasy. You hear your heart pounding through the ringing in your ears, over the music. Then, at once, the music stops.

'Ladies and Gentlemen,' the announcer's voice comes. 'I'm afraid we have a situation. One of Madame Arachne's lovers has escaped the stage. I can only surmise that one of our guests has become sexually aroused. Please, I am referring to the guest in the front row'. Your eyes go wide and you look around frantically for the source of the voice. 'Please! I beg you, do not move. Yes, you, Sir, or Madame. Do not move, please. Please, for the love of God, remain absolutely still.'
You're tucked back in your chair, your legs crossed in front of you. Your hands hang at your side. You feel a tickle at the end of your fingers. Your heart pounding, you glance down.

The announcer shrieks, and you jump in your chair. 'Please! Sir, Madame, do not move, please! For the love of God! Remain absolutely still. Ab-so-lute-ly still, Sir. Or Madame.'

Your hand shakes; there is a big black spider crawling slowly but inexorably over your wrist. You look down and realize that several more spiders are crawling up your pant legs. Your throat closes and your breath stops in an instant as you realize that another spider - the biggest of them all - rests on your lap, stretched languidly across your sexual organs.

'Now, folks, I must insist that everyone in the club remain in their seats; I tell you again, these are the most venomous spiders known to man. Arachnia Venomouso. They can kill a grown human with one bite. We must only hope that Madame Arachne can help us in this situation.'

You turn your eyes toward the stage. Madame Arachne has risen, still covered with her carpet of spiders. Only her feet remain mostly free of the beasts. She moves toward you, slowly, deliberately. She mounts the few stairs leading from the stage to the front row. She approaches you.

'No,' you manage to utter, a strangled sound low in your throat.

'Please, my friend, remain still. Madame Arachne is your only hope now.'

Madame Arachne's eyes flash in the red and white spots from overhead. They are locked in yours. Spiders darken her foreheads like oversized eyebrows. Spiders tangle lumpy in her long dark hair. You look up at her, a silent plea in your eyes.

You smell Madame Arachne's body as she leans toward you, putting one arm around your shoulder. Her naked breasts, covered with furry arachnids, sway close to your face as she bends low against you. You are hunkered down in your chair and the tall Arachne perches on high heels far above you, so that when she bends over you your face is close to her crotch. Two spiders cling to the trimmed dark thatch of her pubic hair. You breathe the sharp scent your own fear mingled with that of her sex. Her fingers find your cheek and gently caress you, as they might a lover.

Madame Arachne plucks a spider from the side of your face. You feel its spiny legs scraping your flesh as it leaves you. Madame Arachne places the animal on her shoulder. She bends lower, lets her hand travel up your thigh. She plucks the spider off your crotch and tucks it against her own. It clings, half-dangling, to her pubic hair. Madame Arachne bends lower and you smell her, stronger, pungent female musk mixed with spices. She takes three spiders in rapid succession from your arm, placing them on her own.

Finally, she crouches low between your legs and lovingly gathers the spiders who have crawled up your leg.

She stands slowly, balanced on high heels between your splayed legs. She takes the spiders from her forehead and cheeks, placing them on the pile gracing her shoulders like a cloak. Madame Arachne bends forward and brings her face close to yours. She kisses you, her lips parting your own and her tongue snaking languidly into your mouth. You taste cigarettes and whiskey on her mouth. As she kisses you, you feel an unexpected scurry of legs from the top of your head, another from your ear, the brush of spiny legs across your cheek. When she draws back, her face is again blotched with spiders - six of them, perhaps, or eight.

Madame Arachne backs slowly away from you, the club's silence suddenly inescapable, like a press of heat all around you. Madame Arachne returns to the stage, and the announcer emits a long, low sigh into the microphone, the cheap electronics crackling and softly whining.

'Thank whatever gods or goddesses you wish, ladies and gentlemen. Our guest is all right. Madame Arachne has saved the day.' Scattered applause erupts from the haunted souls in the back of the club. Shaking, you get up, kick your chair back, hear it fall over and hit the carpeted floor with a thunk. You run your hands over your arms, down your legs; you kick your feet to make sure no creatures go flying. As the music rises and Madame Arachne returns to her dance, you grab your backpack and head for the door, your legs like rubber, your heart still pounding.

Trick. It was all a cheap trick. Some carnival huckster French-quarter trick to tempt future tourists with promises of a sick thrill. I tell the story and my friends all come here, laughing, expecting a pulse-pounding ride of terror. *Fuck them*, you think. *I'm not telling a soul. Never. I'll never speak of it. They can get their free advertising elsewhere, motherfuckers.*

You push through the red velvet curtain, its brush on your skin making you shudder. The doorman draws aside the leather curtain in front, like he knows you're coming.

'Have fun?' he asks.

You shoot him a wicked look, stumbling into the narrow alley and half running on shaking legs toward the voices and stink of Bourbon Street. You've made it halfway down when your knees suddenly give out and you fall against the bare brick wall. Your fear has paralyzed you. You sink to a sitting position, back propped against the wall. Your backpack lays forgotten between your legs. You take a deep breath and go to get up.

Your joints are stiff, hard, immobile. You look back at the doorman, who is puffing his cigarette, looking at you. He perches on his stool, lifts his top hat to you, and grins.

Then you see it. The swelling mound at the webbed crux of your thumb and forefinger. You open your mouth to shout for help. Your neck muscles are paralyzed.

You slump over to the side. 'It was a trick,' you manage to say, your lungs closing, your tongue thick in your mouth. You feel the heat start to hit you, the pulse start to undulate through your body.

You can smell the scents coming from the street: sex, liquor, smoke. But even above all that you can smell what wafts out of the entrance to the club: spiders.

Your joints stiff, you haul yourself to your hands and knees. You try to rise to your feet, but you can't. Instead, you start crawling back toward the club entrance.

The fever is in you now, fiery liquid pulsing through your veins. The scent of Madame Arachne's spiders fills your nostrils, and you feel yourself crawling toward the club. Through piles of garbage, damp pools of piss. You smell it, stronger, now, as the doorman draws back the leather curtain with a grin.

Madame Arachne has come to the door of the club, her body covered in the furry bodies of spiders. You crawl toward her as her frozen, expressionless face twists and for the first time you see her smile. At the doorway, the doorman helps you to your feet and you stumble against the doorjamb. Mistress Arachne holds out one hand; in her palm is a spider.

You take the spider from her and place it on the back of your hand. This time you feel it when the mandibles penetrate your flesh. With its bite, the fire courses through you; dropping to your knees again, and Mistress Aarachne, nude except for her cloak of lovers, turns and beckons you into the club.

You crawl into the darkness, moaning.

Bonnie Ruberg

PRINCESS PEACH THE PORN STAR
POWER DYNAMICS IN EROTIC VIDEO GAME FAN FICTION

What Is Erotic Video Game Fan Fiction?

Life can be tough for video game fans, or any fans for that matter. They love games with all their endearingly dorky hearts, and though they can play them over and over, they can never really interact with them, or communicate their affection to them – at least not in way people can do with one another. Instead, pressed for an outlet to express their fandom, dedicated game lovers often turn to creating. They use the characters and stories from their favorite titles and incorporate them into drawings, crafts, and even stories. It's these stories, called fan fiction that we want to talk about today. In particular, we'll be looking at something that comes up time and time again in fan fiction: sex.

Before diving headlong into hot Mario-on-Peach action, let's start with the larger issue of video games and sex and work our way back. Like any creative medium, games have plenty of potential for sexiness, and lots of examples of that have already cropped up. Most of them fit within three categories. We have sexy video game characters, like the infamously busty Lara Croft, who bring eroticism into gaming through character design. We have the brief and often controversial blips of sex that occasionally sneak into mainstream games, like the interactive intercourse mini-game in Grand Theft Auto: San Andreas known as Hot Coffee. And we have games dedicated entirely to sex, such as adult titles like the online simulation game Virtually Jenna.

However, we don't intend to talk about any of those things. Instead, our discussion centers around games that have, to the untrained eye, no sexual content whatsoever. These classic games, like those in The Legend of Zelda or Mario Bros. series, come in multiple installments, which have been released literally over decades, and feature recurring characters that video game fans have grown attached to. In such titles a beautiful princess may need saving, but aside from a few accidental up skirt shots on windy fictional days (why do female royalty always wear such large undergarments?) those in search of steamy pixilated action should look elsewhere.

That is, until fans enter the picture. Despite the stereotypes about gamers, they come in all genders, ages, and levels of dedication. Similarly, each fan has a slightly different way of expressing his affection for a beloved game. Some spend time on fan sites and forums. Others replay old titles. However, a surprisingly large subsection of followers respond to the games they love - those works of interactive art - by creating something artistic themselves.

The online community of fan fiction writers has been going strong since the creation of games themselves. Actually, it started years before with Sci-Fi cult classics of 1960's TV, like the original Star Trek. Fan fiction, for those among who've never accidentally stumbled across its sultry adventures while googling for something completely 'innocent,' incorporates characters and situations into short stories. And a portion of those stories turn into hardcore sex-fests... Well, prepare to be impressed.

No One Takes This Stuff Seriously, Right?

Writing stories may seem like a pretty elaborate – and therefore rare - way to express video game fandom, but in fact tons of websites exist for this sort of thing. Fanfiction.net has published literally hundreds of thousands of such tales. 50,820 stories have been created in homage to Final Fantasy games on that site alone. Also popular: Zelda games (12,444), Sonic the Hedgehog (12,156), and Kingdom Hearts (37, 341). Asexual classics consistently win out over tantalizing new hits.

It's difficult to nail down exact demographics on who writes video game fan fiction, but the internet community generally assumes the numbers skew toward younger gamers – and potentially a less male-dominated crowd than in the rest of the games community, since girls generally tend to gravitate toward writing as a means of self-expression. The quality of writing in these stories can range from decent to decently low. Spelling and grammar mistakes come up as commonly as video game references themselves. Sure, this makes some fan fiction laughable, but it's also a testament to how many young creative types it inspires to start writing.

Browsing through an archive of video game fan fiction – as with any type of fan fic –what stands out more than the spelling errors is the sex. In the Zelda category of FanFiction.net, for example, we instantly run across a story called 'Twilight Love.' A quick, curious skim tells us that in this tale Link and Midna (the female imp from Twilight Princess) share a romantic dinner and eventually make sweet love. In the Sonic category we're confronted with 'Nectar Sweet,' in which Tails hits puberty and starts fantasizing about multi-tailed sex. The Sonic category alone holds 600+ stories rated Mature for sexual content.

If only a handful of such stories existed out there on the internet, we could write them off as offshoots of Rule #34, which UrbanDictionary.com defines as a 'gaccepted internet rule that states that pornography or sexually related material exists for any conceivable subject.' In summary: if it exists, there's porn of it. However, there are tons of these things, and together they make a pattern – or even a phenomenon. They show us a widespread way in which fans are reacting to a digital entertainment medium. That is, they react with sexual energy. Instead of dealing with abstracts though, let's hear a few excerpts from the stories we've mentioned:

'Twilight Love'

Link has just come back from running in the woods. Midna makes him dinner. In case you're wondering, no, this doesn't fit in anywhere with the actual Twilight Princess plot:
Link stood, walked over, and kissed Midna gently. Midna hesitated for a second before kissing back. Link picked up the little imp and carried her to her room, setting her on the bed. She looked up at him and smiled. He was starting to get hard and she could tell. Quickly, she undid his belt and let his breeches drop. He kissed her lightly and smiled. He rubbed her thigh and smiled when he felt, to his surprise, she was wet. She moaned and looked at his engorged penis. She had never been one to be on her knees, but that didn't stop her. She got to the floor and licked his penis gently, before taking it in her mouth.

'Nectar Sweet'

This is how this story opens. Remember, we're talking about a small, fuzzy, orange fox:
She led me into a dark room, only lit by a single candle. 'What are we doing here?' I asked, knowing the answer already. Her response was a wink, and a shut of the door. My stomach was in knots, and I could barely keep from passing out. Her facial features showed an immense amount of lust, which didn't help cease my stomach. 'Tails…' she whispered, slowly stepping closer to me. My breathing intensified, and her smile deepened by every stuttered sentence I attempted to speak. I started to open my mouth to speak, but she put a finger to my lips, forcing me to stop. She put her arms around my head, and pulled me towards her, into a soft kiss. We stood there for a few minutes, hours? Who knows? We were lost in our own worlds, but just as suddenly as it started, she pulled away. 'Cosmo, we-' I was silenced by her action of removing the blouse she wore, revealing her most private areas.

Clearly we're dealing with some pretty hardcore stuff. From a cultural analysis standpoint, our goal is to ask, 'Why?' Why do these stories employ graphic sex? Why do they employ sex at all? And why do they inject sexual imagery into games that appear otherwise completely asexual?

Why Do Fans Write These Stories?

We can boil down all those questions into one: why do people write erotic video game fan fiction? Ask gamers, game professionals, or even yourself, and you're bound to come up with a smattering of the following speculative explanations:

- Fans have crushes on their favorite characters and so naturally want to see them – or at least imagine them – naked and getting it on. What's so complicated? Our problem with this argument: it oversimplifies things, rendering the sexual nature of fan fiction almost irrelevant.

- These stories are written by hormone-driven gamers who recently hit puberty and don't know where else to channel their sexual energy. Everything they touch turns into sex, so why not their fan fiction? Our problem with this one: it's still too simple. It implies that the writers of fan fiction have no choice but to write erotica, no agency over their content.

- Erotic fan fiction offers a venue for younger, inexperienced fans to approach sex in a non-threatening way. While sex in the abstract may seem daunting, these stories give them a chance to explore the acts they've been hearing/thinking about with characters that feel familiar and safe. Our problem: actually, this theory makes a lot of sense. It just doesn't happen to be the answer we're looking for.

- Fans write about what lies outside the realm of the official games. They're beloved G-rated characters will never have sex on-screen, and so it's that they need to pen themselves. Adventure they can already get in abundance from the games proper. In series like The Legend of Zelda or Mario Bros, for example, we often see a male hero save a princess – but we never seem them consummate their victory. Writers want to envision what happens after the game ends, after the tape stops rolling. This is also an interesting and valid theory, but for the sake of argument the one we're after is…

- There's something innately sexual about the act of writing video game fan fiction, which explains why so many of these stories turn erotic.

Bingo! That's the answer we're looking for. But what the heck does it mean?

Taking Back Control and Turning It Sexy

First, let's think about the idea of control. Control plays a very particular role in video games, because they operate on the premise of interactivity. In theory, the player has control over the situation. He can tell his character when to move, whom to shoot, whom to kiss – if such a thing is built into the game. He has, it would appear, power.

At the same time, most video games give players very little control over their actions in a larger sense: where they can go, what sort of actions they can perform, and what they need to do to progress. Items must be collected in a certain order; a terrain can only be traversed in a certain way. Though the player is teased with tantalizing hints of empowerment, he remains largely constrained.

That taste of power, while unconsciously frustrating players, also creates tension and gives video game fans they're artistic spark. Really, what they're doing when they write fan fiction is taking back the control that was so temptingly dangled in front of them before. Instead of playing along with the games' creators, they make well-known characters their own. Suddenly the objects of their affection bow to their will. Fans tell them where to move, what to do.

And that control is sexy.

Because it creates a power dynamic, the act of writing fan fiction is surrounded by tension. That tension, in combination with the newly found power of the writer, sets a fan in a sexually heightened state before he even puts his fingers to the keys. Once he does, it's no wonder that what he produces is sexual. As the one in control, he has free reign to assert his creative desires.

If we follow this argument through – and, granted it's not without its flaws – it explains the sheer amount of erotic (vs. non-sexual) video game fan fiction out there. Writers may be young and horny, some video game characters may be sexualized, but more importantly the act of taking artistic dominance, of making characters your own, is by nature a type of sexual conquest.

So what's next? Now that more and more video games explicitly dedicated to sex are hitting the online market, maybe we'll see a new type of dominance – the kind that removes sex from the picture. Instead of putting friendly hedgehogs and plump plumbers in erotic situations, perhaps we'll encounter fan fiction for games like the sex sim Virtually Jenna that strips characters of their erotic edge. How banal, and yet how bizarre, who these stories sound? 'Today Jenna did her laundry. She used plenty of fabric softener and then folded each sock individually. The end.'

Rudy Rucker

RAPTURE IN SPACE

Denny Blevins was a dreamer who didn't like to think. Drugs and no job put his head in just the right place for this. If at all possible, he liked to get wired and spend the day lying on his rooming house mattress and looking out the window at the sky. On clear days he could watch his eyes' phosphenes against the bright blue; and on cloudy days he'd dig the clouds' drifty motions and boiling edges. One day he realized his window-dirt was like a constant noise-hum in the system, so he knocked out the pane that he usually looked through. The sky was even better then, and when it rained he could watch the drops coming in. At night he might watch the stars, or he might get up and roam the city streets for deals.

His Dad, whom he hadn't seen in several years, died that April. Denny flew out to the funeral. His big brother Allen was there, with Dad's insurance money. Turned out they got $15K apiece.

'Don't squander it, Denny,' said Allen, who was an English teacher. 'Time's winged chariot for no man waits! You're getting older and it's time you found a career. Go to school and learn something. Or buy into a trade. Do something to make Dad's soul proud.'

'I will,' said Denny, feeling defensive. Instead of talking in clear he used the new cyberslang. 'I'll get so cashy and so starry so zip you won't believe it, Allen. I'll get a tunebot, start a motion, and cut a choicey vid. Denny in the Clouds with Clouds. Untense, bro, I've got plex ideas.'

When Denny got back to his room he got a new sound system and a self-playing electric guitar. And scored a lot of dope and food-packs. The days went by; the money dwindled to $9K. Early in June the phone rang.

'Hello, Denny Blevins?' The voice was false and crackly.

'Yes!' Denny was glad to get a call from someone besides Allen. It seemed like lately Allen was constantly calling him up to nag.

'Welcome to the future. I am Phil, a phonebot cybersystem designed to contact consumer prospects. I would like to tell about the on-line possibilities open to you. Shall I continue?'

'Yes,' said Denny.

It turned out the 'Phil the phonebot' was a kind of computerized phone salesman. The phonebot was selling phonebots which you, the consumer prospect, could use to sell phonebots to others. It was - though Denny didn't realize this - a classic Ponzi pyramid scheme, like a chain letter, or like those companies which sell people franchises to sell franchises to sell franchises to sell...

The phonebot had a certain amount of interactivity. It asked a few yes/no questions; and whenever Denny burst in with some comment, it would pause, say, 'That's right, Denny! But listen to the rest!' and continue. Denny was pleased to hear his name so often. Alone in his room, week after week, he'd been feeling his reality fade. Writing original songs for the guitar was harder than he'd expected. It would be nice to have a robot friend. At the end, when Phil asked for his verdict, Denny said, 'Okay, Phil, I want you. Come to my rooming-house tomorrow and I'll have the money.'

The phonebot was not the arm-waving clanker that Denny, in his ignorance, had imagined. It was, rather, a flat metal box that plugged right into the wall phone-socket. The box had a slot for an electronic directory, and a speaker for talking to its owner. It told Denny he could call it Phil; all the phonebots were named Phil. The basic phonebot sales spiel was stored in the Phil's memory, though you could change the patter if you wanted. You could, indeed, use the phonebot to sell things other than phonebots.

The standard sales pitch lasted five minutes, and one minute was allotted to the consumer's responses. If everyone answered, listened, and responded, the phonebot could process ten prospects per hour, and one hundred twenty in a 9 A. M. – 9 P. M. day! The whole system cost nine thousand dollars, though as soon as you bought one and joined the pyramid, you could get more of them for six. Three thousand dollars profit for each phonebot your phonebot could sell! If you sold, say, one a day, you'd make better than $100K a year!

The electronic directory held all the names and numbers in the city; and each morning it would ask Denny who he wanted to try today. He could select the numbers on the day's calling list on the basis of neighborhood, last name, family size, type of business and so on.

The first day, Denny picked a middle-class suburb and told Phil to call all the childless married couples there. Young folks looking for an opportunity! Denny set the speaker so he could listen to people's responses. It was not encouraging.
'click'

'No ... *click*'

'*click*'

'This kilp ought to be illegal ... *click*'

'*click*'

'Get a job, you bizzy dook ... *click*'

'Of all the ... *click*'

'Again? *click*'

Most people hung up so fast that Phil was able to make some thousand unsuccessful contacts in less than ten hours. Only seven people listened through the whole message and left comments at the end; and six of these people seemed to be bedridden or crazy. The seventh had a phonebot she wanted to sell cheap.

Denny tried different phoning strategies - rich people, poor people, people with two sevens in their phone number, and so on. He tried different kinds of sales pitches - bossy ones, ingratiating ones, curt ones, ethnic-accent ones, etc. He made up a sales pitch that offered businesses the chance to rent Phil to do phone advertising for them.

Nothing worked. It got to be depressing sitting in his room watching Phil fail - it was like having Willy Loman for his room-mate. The machine made little noises, and unless Denny took a *lot* of dope, he had trouble relaxing out into the sky. The empty food-packs stank.

Two more weeks, and all the money, food and dope were gone. Right after he did the last of the dope, Denny recorded a final sales-pitch:

'Uh ... hi. This is Phil the prophet at 1801 Eye Street. I eye I ... I'm out of money and I'd rather not have to ... uh ... leave my room. You send me money for ... uh ... food and I'll give God your name. Dope's rail, too.'

Phil ran that on random numbers for two days with no success. Denny came down into deep hunger. Involuntary detox. If his Dad had left much more money, Denny might have died, holed up in that room. Good old Dad. He trembled out into the street and got a job working counter in a Greek coffee shop called the KoDo. It was okay; there was plenty of food, and he didn't have to watch Phil panhandling.

As Denny's strength and sanity came back, he remembered sex. But he didn't know any girls. He took Phil off panhandling and put him onto propositioning numbers in the young working-girl neighborhoods.

'Hi, are you a woman? I'm Phil, sleek robot for a whippy young man who's ready to get under. Make a guess and he'll mess. Leave your number and state your need; he's fuff-looking and into sleaze.'

This message worked surprisingly well. The day after he started it up, Denny came home to find four enthusiastic responses stored on Phil's chips. Two of the responses seemed to be from men, and one of the women's voice sounded old ... *really* old. The fourth response was from 'Silke.'

'Hi, desperado, this is Silke. I like your machine. Call me.'

Phil had Silke's number stored, of course, so Denny called her right up. Feeling shy, he talked through Phil, using the machine as voder to make his voice sound weird. After all, Phil was the one who knew her.

'Hello?' Cute, eager, practical, strange.

'Silke? This is Phil. Denny's talking through me. You want to interface?'

'Like where?'

'My room?'

'Is it small? It sounds like your room is small. I like small rooms.'

'You got it. 1801 Eye Street, Denny'll be in front of the building.'

'What do you look like, Denny?'

'Tall, thin, teeth when I grin, which is lots. My hair's peroxide blond on top. I'll wear my X-shirt.'

'Me too. See you in an hour.'

Denny put on his X-shirt - a T-shirt with a big silk screen picture of his genitalia - and raced down to the KoDo to beg Spiros, the boss, for an advance on his wages.

'Please, Spiros, I got a date.'

The shop was almost empty, and Spiros was sitting at the counter watching a payvid porno show on his pocket TV. He glanced over at Denny all decked out in his X-shirt, and pulled two fifties out of his pocket.

'Let me know how she comes.'

Denny spent one fifty on two Fiesta food-packs and some wine: the other fifty went for a capsule of snap-crystals from a street vendor. He was back in front of his rooming-house in plenty of time. Ten minutes, and there came Silke, with a great big pink crotch-shot printed onto her T-shirt. She looked giga good.

For the first instant they stood looking at each other's X-shirts, and then they shook hands.

'I'm Denny Blevins. I've got some food and wine and snap here, if you want to go up.' Denny was indeed tall and thin, and toothy when he grinned. His mouth was very wide. His hair was long and dark in back, and short and blond on top. He wore red rhinestone earrings, his semierect X-shirt, tight black plastic pants, and fake leopard fur shoes. His arms were muscular and veiny, and he moved them a lot when he talked.

'Go up and get under,' smiled Silke. She was medium height, and wore her straw-like black hair in a bouffant. She had fine, hard features. She'd appliquéd pictures of monster eyes to her eyelids, and she wore white day-glo lipstick. Beneath her sopping wet X-shirt, she wore a tight, silvered jumpsuit with cutouts. On her feet she wore roller skates with lights in the wheels.

'Oxo,' said Denny.

'Wow,' said Silke.

Up in the room they got to know each other. Denny showed Silke his phonebot and his sound system, pretended to start to play his guitar and to then decide not to, and told about some of the weird things he'd seen in the sky, looking out that broken pane. Silke, as it turned out, was a payvid sex dancer from West Virginia. She talked mostly in clear, but she was smart, and she liked to get wild, but only with the right kind of guy. Sex dancer didn't mean hooker and she was, she assured Denny, clean. She had a big dream she wasn't quite willing to tell him yet.

'Come on,' he urged, popping the autowave food-packs open. 'Decode.'

'Ah, I don't know, Denny. You might think I'm skanky.'

They sat side by side on Denny's mattress and ate the pasty food with the packed in plastic spoons. It was good. It was good to have another person in the room here.

'Silke,' said Denny when they finished eating, 'I'd been thinking Phil was kilp. Dook null. But if he got you here it was worth it. Seems I just need tech to relate, you wave?'

Silke threw the empty food trays on the floor and gave Denny a big kiss. They went ahead and fuffed. It seemed like it had been a while for both of them. Skin all over, soft, warm, skin, touch, kiss, lick, smell, good.

Afterwards, Denny opened the capsule of snap and they split it. You put the stuff on your tongue, it sputtered and popped, and you breathed in the freebase fumes. Fab rush. Out through the empty window pane they could see the moon and two stars stronger than the city lights.

'Out there,' said Silke, her voice fast and shaky from the snap. 'That's my dream. If we hurry, Denny, we can be the first people to have sex in space. They'd remember us forever. I've been thinking about it, and there was always missing links, but you and Phil are it. We'll get in the shuttlebox - it's a room like this - and go up. We get up there and make videos of us getting under, and - this is my new flash - we use Phil to sell the vids to pay for the trip. You wave?'

Denny's long, maniacal smile curled across his face. The snap was still crackling on his tongue. 'Stuzzadelic! Nobody's fuffed in space yet? None of those gawks who've used the shuttlebox?'

'They might have, but not for the record. But if we scurry we'll be the famous first forever. We'll be starry.'

'Oxo, Silke.' Denny's voice rose with excitement. 'Are you there, Phil?'

'Yes, Denny.'

'Got a new pitch. In clear.'

'Proceed.'

'Hi, this is Denny.' He nudged the naked girl next to him.

'And this is Silke.'

'We're doing a live fuff-vid we'd like to show you.'

'It's called *Rapture in Space*. It's the very first X-rated love film from outer space.'

'Zero gravity,' said Denny, reaching over to whang on his guitar.

'Endless fun.'

'Mindless pleasure.' *Whang.*

'Out near the sun.' Silke nuzzled his neck and moaned stagily. 'Oh, Denny, oh, darling, it's ...'

'*RAPTURE IN SPACE*! Satisfaction guaranteed. This is bound to be a collector's item; the very first live sex video from space. A full ninety minutes of unbelievable null-gee action, with great Mother Earth in the background, tune in for only fifty ...'

'More, Denny,' wailed Silke, who was now grinding herself against him with some urgency. 'More!'

Whang. 'Only one hundred dollars, and going up fast. To order, simply leave your card number after the beep.'

'*Beep!*'

Phil got to work the next morning, calling numbers of businesses where lots of men worked. The orders poured in. Lacking a business-front by which to cash the credit orders, Denny enlisted Spiros, who quickly set up KoDo Space Rapture Enterprises. For managing the business, Spiros only wanted 15% and some preliminary tapes of Denny and Silke in action. For another 45%, Silke's porno payvid employers - an outfit known as XVID - stood ready to distribute the show. Dreaming of this day, Silke had already bought her own cameras. She and Denny practiced a lot, getting their moves down. Spiros agreed that the rushes looked good. Denny went ahead and reserved the shuttlebox for a trip in mid July.

The shuttlebox was a small passenger module that could be loaded into the space shuttle for one of its weekly trips up to orbit and back. A trip for two cost $100K. Denny bought electronic directories for cities across the country, and set Phil to working twenty hours a day. He averaged fifty sales a day, and by launch time, Silke and Denny had enough to pay everyone off, and then some.

But this was just the beginning. Three days before the launch, the news services picked up on the *Rapture in Space* plan, and everything went crazy. There was no way for a cheap box like Phil to process the orders anymore. Denny and Silke had to give XVID anther 15% of the action, and let them handle the tens of thousands of orders. It was projected that *Rapture in Space* would pull an audience share of 7% - which is a lot of people. Even more money came in the form of fat contracts for two product endorsements: SPACE RAPTURE, the cosmic eroscent for high-flyers, and RAPT SHIELD, an antiviral lotion for use by sexual adventurers. XVID and the advertisers privately wished that Denny and Silke were a bit more ... *upscale* looking, but they were the two who had the tiger by the tail.

Inevitably, some of the Christian Party congressmen tried to have Denny and Silke enjoined from making an XVID broadcast from aboard the space shuttle which was, after all, government property. But for 5% of the gross, a fast thinking lawyer was able to convince a hastily convened Federal court that, insofar as *Rapture in Space* was being codecast to the XVID dish and cabled thence only to paying subscribers, the show was a form of constitutionally protected free speech, in no way essentially different from a live-sex show in a private club.

So the great day came. Naked save for a drenching of Space Rapture eroscent, Silke and Denny waved goodbye and stepped into their shuttle-box. It was shaped like a two-meter-thick letter D, with a rounded floor, and with a big picture window set into the flat ceiling. A crane loaded the shuttlebox into the bay of the space shuttle along with some satellites, missiles, building materials, etc. A worker dogged all the stuff down, and then the baydoors closed. Silke and Denny wedged themselves down into their puttylike floor. Blast off - roar, shudder, push, clunk, roar some more.

Then they were floating. The baydoors swung open, and the astronauts got to work with their retractable arms and space tools. Silke and Denny were busy, too. They set up the cameras, and got their little antenna locked in on the XVID dish. They started broadcasting right away - some of the *Rapture in Space* subscribers had signed up for the whole live protocols in addition to the ninety minute show that Silke and Denny were scheduled to put on in...

'Only half an hour, Denny,' said Silke. 'Only thirty minutes till we go on.' She was crouched over the sink, douching, and vacuuming the water back up. As fate would have it, she was menstruating. She hadn't warned anyone about it.

Denny felt cold and sick to his stomach. XVID had scheduled their show right after take-off because otherwise - with all the news going on - people might forget about it. But right now he didn't fell like fuffing at all, let alone getting under. Every time he touched something, or even breathed, his whole body moved.

'All clean now,' sang Silke. 'No one can tell, not even you.'

There was a rapping on their window - one of the astronauts, a jolly jock woman named Judy. She grinned through her helmet and gave them a high sign. The astronauts thought the *Rapture in Space* show was a great idea; it would make people think about them in new, more interesting ways.

'I talked to Judy before the launch,' said Silke, waving back. 'She said to watch out for the rebound.' She floated to Denny and began fondling him. 'Ten minutes, starman.'

Outside the window, Judy was a shiny wad against Earth's great marbled curve. *The clouds,* Denny realized, *I'm seeing the clouds from on top.* His genitals were warming to Silke's touch. He tongued a snap crystal out of a crack between his teeth and bit it open. Inhale. The clouds. Silke's touch. He was hard, thank God, he was hard. This was going to be all right.

The cameras made a noise to signal the start of the main transmission, and Denny decided to start by planting a kiss on Silke's mouth. He bumped her shoulder and she started to drift away. She tightened her grip on his penis and led him along after her. It hurt, but not too unpleasantly. She landed on her back, on the padded floor, and guided Denny right into her vagina. Smooth and warm. Good. Denny pushed into her and ... *rebound.*

He flew, rapidly and buttocks first, up to the window. He had hold of Silke's armpit and she came with him. She got her mouth over his penis for a second, which was good, but then her body spun around, and she slid toothrakingly off him, which was very bad.

Trying to hold a smile, Denny stole a look at the clock. Three minutes. *Rapture in Space* had been on for three minutes now. Eighty-seven minutes to go.

It was another bruising half hour or so until Denny and Silke began to get the hang of spacefuffing. And then it was fun. For a long time they hung in midair, with Denny in Silke, and Silke's legs around his waist, just gently jogging, but moaning and throwing their heads around for the camera. Actually, the more they hammed it up, the better it felt. Autosuggestion.

Denny stared and stared at the clouds to keep from coming, but finally he had to pull out for a rest. To keep things going they did rebounds for awhile. Silke would lie spread-eagled on the floor, and Denny would kind of leap down on her; both of them adjusting their pelvises for a bullseye. She'd sink into the cushions, then rebound them both up. It got better and better. Silke curled up into a ball and impaled herself on Denny's shaft. He wedged himself against the wall with his feet and one hand and used his other hand to spin her around and around, bobbin on his spindle. Denny lay on the floor and Silke did leaps onto him. They kissed and licked each other all over, and from every angle. The time was almost up.

For the finale, they went back to midair fuffing; arms and legs wrapped around each other; one camera aimed at their faces, and one camera aimed at their genitalia. They hit a rhythm where they always pushed just as hard as each other and it action/reaction cancelled out, hard and harder, with big Earth out the window, yes, the air full of their smells, yes, the only sound the sound of their ragged breathing, yes, now NOW AAAHHHHHHH!!!!

Denny kind of fainted there, and forgot to slide out for the come-shot. Silke went blank, too, and they just floated, linked like puzzle pieces for five or ten minutes. It made a great finale for the *Rapture in Space* show, really much more convincing than the standard sperm spurt.

Two days later, and they were back on Earth, with the difference that they were now, as Denny had hoped, cashy and starry. People recognized them everywhere, and looked at them funny, often asking for a date. They did some interviews, some more endorsements and they got an XVID contract to host a monthly spacefuff variety show.

Things were going really good until Denny got a tumor.

'It's a dooky little kilp down in my bag,' he complained to Silke. 'Feel it.'

Sure enough, there was a one-centimeter lump in Denny's scrotum. Silke wanted him to see a doctor, but he kept stalling. He was afraid they'd run a blood test and get on his case about drugs. Some things were still illegal.

A month went by and the lump was the size of an orange.

'It's so gawky you can see it through my pants,' complained Denny. 'It's giga ouch and I can't cut a vid this way.'

But he still wouldn't go to the doctor. What with all the snap he could buy, and with his new cloud telescope, Denny didn't notice what was going on in his body most of the time. He was happy to miss the next few XVID dates. Silke hosted them alone.

Three more months and the lump was like a small watermelon. When Denny came down one time and noticed that the tumor was moving he really got worried

'Silke! It's alive! The thing in my bag is alive! Aaauuugh!'

Silke paid a doctor two thousand dollars to come to their apartment. The doctor was a bald, dignified man with a white beard. He examined Denny's scrotum for a long time, feeling, listening, and watching the tumor's occasional twitches. Finally he pulled the covers back over Denny and sat down. He regarded Silke and Denny in silence for quite some time. 'Decode!' demanded Denny. 'What the kilp we got running here?'

'You're pregnant,' said the doctor. 'Four months into it, I'd say.'

The quickening fetus gave another kick and Denny groaned. He knew it was true. 'But how?'

The doctor steepled his fingers. 'I ... I saw *Rapture in Space*. There were certain signs to indicate that your uh partner was menstruating?'

'Check.'

'Menstruation, as you must know, involves the discharge of the unfertilized ovum along with some discarded uterine tissues. I would speculate that after your ejaculation the ovum became wedged in your meatus. It is conceivable that under weightless conditions the sperm's flagell3/4 could have driven the now-fertilized ovum up into your vas deferens. The ovum implanted itself in the blood-rich tissues there and developed into a fetus.'

'I want an abortion.'

'No!' protested Silke. 'That's our baby, Denny. You're already almost half done carrying it. It'll be lovely for us ... and just think of the publicity!'

'Uh ...' said Denny, reaching for his bag of dope.

'No more drugs,' said the doctor, snatching the bag. 'Except for the ones I give you.' He broke into a broad, excited smile. 'This will make medical history.'

And indeed it did. The doctor designed Denny a kind of pouch in which he could carry his pregnant scrotum, and Denny made a number of video appearances, not all of them X-rated. He spoke on the changing roles of the sexes, and he counted the days till delivery. In the public's mind, Denny became the symbol of a new recombining of sex with life and love. In Denny's own mind, he finally became a productive and worthwhile person. The baby was a flawless girl, delivered by a modified Caesarian section.

Sex was never the same again.

I wrote this story in Lynchburg, Virginia, in the fall of 1984. It appeared in an avant-garde SF anthology edited by Peter Lamborn Wilson, Robert Anton Wilson, and me, Semiotext[e] SF, *(New York: Autonomedia, 1989). The story also appears in my personal anthology,* Gnarl *(New York: Running Press, 2000).*

Mae Saslaw

PERFORMANCE, IDENTITY, AND SUBVERSION
SEX AND GENDER IN THE AGE OF SOCIAL NETWORKING

There are two forces at work on the Internet, two intellectual armies at a virtual impasse: There's us, the people who looked at this thing when it first happened and thought, 'Oh good, finally some democracy!' or better yet, 'Oh good, finally some anarchy!' We love that everything is possible and everything worthwhile is free, we see zeros and ones as art and philosophy and not as commodities. Then there's them, who have been producing technology for the purposes of commodifying un-commodified markets, who sell us new electronics so they can sell us more electronics. They use the Internet to make buying easier, to remove the sense of exchanging money for goods and services that our capitalist system has been based on for hundreds of years so that we start to think we're not exchanging money at all - and yet the money is gone just the same. This is a discussion of how the virtual impasse makes its way into our sexuality, and how different people react to the new possibilities. We can either use the Internet to keep doing what we've been doing, and just do it with more people more of the time (that's them); or we can use it to completely change the way things are done altogether.

The big deal about sexuality on the Internet - the reason we must deal with it - is that we don't need bodies for it. We can be whatever we want; we can express those parts of our identities that diverge from biology. The obvious example might be transgendered people: If you want to live with a different gender then the one you were born with, it's incredibly easy to do on the Internet: you just choose your preferred gender from the drop-down menu, and that's it. You're a man now, or a woman, as far as the Internet is concerned. But it goes farther than that: What if you don't identify with your race, your body type, your age, your socioeconomic position, your nationality? Just present yourself as whatever you want. Pick your own priorities. It's almost as if all of the problems of biology, binarisms, and hegemony evaporate right away. But aren't the same structures still present on the Internet? We can't simply introduce fluid identity absent biological predicates and call it a day. To quote Judith Butler: '…if the model of a more diffuse and antigenital sexuality serves as the singular, oppositional alternative to the hegemonic structure of sexuality, to what extent is that binary relation fated to reproduce itself endlessly?'[1] In other words, we just get a new binary: a subversive paradigm against a hegemonic one. The next thing Butler asks is 'What possibilities exist for the disruption of the oppositional binary itself?'[2] We have to use that subversive paradigm to dismantle the old one, and, conveniently, identities not predicated on biology raises questions about that hegemonic structure of sexuality. Our awareness of what goes on in the dark corners of the Internet - and the not-so-dark corners - is already influencing the way people think about real life sexuality.

But returning to what performance constitutes, let's imagine an individual who goes so far as to identify with the body she has in real life. Let's call her Judy. When Judy gets up in the morning and gets ready to go outside, she's got some choices to make, and a lot of these are political choices, whether she knows it or not. In fact, they might all be political choices. She has some options, and all of them tell us something about her identity. The importance she places on each of these choices also has a lot to do with her identity. They're all part of her identity in that she takes into account how she feels about her gender and sexuality and either uses these choices to reflect her feelings, or uses these choices to make it seem as though she feels otherwise. She's in the system, and either she likes it or she doesn't like it, either she tells us or she doesn't tell us. So now, let's watch Judy make a MySpace profile. Now Judy gets to control everything, and she's got way more choices. She can talk to us however she wants to talk to us, and there's nothing we can do about it, because Judy is pretty sure that we'll either like what we see or we'll move on. These are all still political choices, and we'll probably interpret them as such, but the point is that Judy has them. Judy's MySpace profile is all performance; there's no chance of our seeing information she doesn't want us to see, and so on. The profile is orchestrated, calculated, and while we can scan for subtext or missteps all we want, we know that Judy probably put more intent behind every word and image we see than she put in her morning routine.

[1] Judith Butler: Gender Trouble. New York: Routledge 2006. p. 37.
[2] Ibid.

Here comes the really tricky question: Which one is the 'real' Judy, and why? She could be lying on her profile! Maybe we even assume that she is. But now we've got to ask what constitutes lying, what it means to misrepresent, and what it means to represent. We've got to ask if the profile is a representation at all, or if it's a presentation. The main problem is that we probably want to give more credence to the Judy we saw getting ready in the morning, since we're hung up on this real life interface. But what if Judy doesn't give more credence to that version of herself? Is the still lying? Do we think she's lying? I think we're likely to look at the discrepancies between the two Judys and interpret them as Judy taking some kind of poetic license with her identity. We make these little allowances; we forgive Judy's carving out space for something that we still read as her personal mythology, and not as her real identity. But for Judy, her new performance isn't about just presenting an alternate version of herself; it's about actually becoming an identity she doesn't otherwise get to be in a world - the Internet - that might in fact be just as legitimate as, and maybe more so than, real life. Because the conversation isn't one-sided, because Judy doesn't just make her profile and stop there, because the Internet is way, way bigger than the interior monologue Judy's had about her identity her whole life. Because on the Internet, there are other people, and there is also sex. The creation of identity in this new environment also functions slightly differently. It's as if the Internet is one big mirror stage. We're forming an ideal, and then becoming it. And if the Internet is a mirror stage, we get rid of the problem of identities not being real.

What bothers us about the discrepancy between the two Judys? Judy isn't even an extreme example; we've cast her as someone whose online performance just about matches up with her real life performance. What concerns us is that we're not sure, looking at her profile, what she'd be like in real life. This makes us very, very anxious. This is where, for us, it's a problem that she might be lying, because we're very concerned with her real life identity, since this is the one we think we might someday encounter sexually. And as far as real life sexual encounters go, the body is still very important. More important than that is the way Judy fits in to the hegemonic sexual paradigm. Going back to Butler, one of her main arguments in Gender Trouble is that identity and in fact personhood are reliant on the paradigm. She writes, and the quotes are hers, 'Inasmuch as 'identity' is assured through the stabilizing concepts of sex, gender, and sexuality, the very notion of 'the person' is called into question by the cultural emergence of those 'incoherent' or 'discontinuous' gendered beings who appear to be persons but who fail to conform to the gendered norms of cultural intelligibility by which persons are defined.' And then, later on: 'the spectres of discontinuity and incoherence, themselves thinkable only in relation to existing norms of continuity and coherence, are constantly prohibited and produced by the very laws that seek to establish causal or expressive lines of connection among biological sex, culturally constituted genders, and the 'expression' or 'effect' of both in the manifestation of sexual desire through sexual practice.'[3] In other words, we might not even be able to read people who don't fit the paradigm as people at all. People become people through their assimilation into the normative structure, and the cues we have for reading identity are only constructed via the individual's participation in the patriarchal system. All these systems of normativity exist because the powers that be have a vested interest in regulating sexuality; we need to be productive members of society, and we need to be reproductive members of society, and, supposedly, we can't do that if our identities fall outside the scope of 'coherence.' Now, what

[3] Ibid., p. 23.

constitutes legible sexuality on the Internet? Do we use the same criteria? Do we apply the same confines, the same restrictions the patriarchy has been so successful at getting us to apply in real life? Here's where the virtual impasse comes back: Obviously, they want us to see Internet sexuality as analogous to real life sexuality, and to apply the same standards. Our goal must be to make entirely new definitions of sexuality.

I see Internet dating as their primary tool for keeping online sexuality in check, for keeping it in line with the existing constructions. Internet dating helps foster the anxiety we have over that slippery disparity between real life identity and online identity, because it's taking everything we know about sex, about reading people and meeting people, and just translating the current practices into a new interface - just like they did with shopping. Paid Internet dating preys upon the freedom of expression and presentation that is unique to the online world. While the matchmaking trade is anything but new, the Internet provides an opportunity to expand matchmaking into un-commodified markets. Desire gets sold in private without the individual ever having to confront another human being, and it's the sense of privacy that counts. Online daters love their privacy and anonymity, and hate the privacy and anonymity of everyone else. So they'll pay and pay to see more of their peers, but the anxiety never goes away, fueled by the sense that the dater, having paid up, is entitled to the 'honesty' of everybody else. And this honesty becomes so much more important in the paid online dating community because, obviously, the users are essentially paying for sex, or the promise of sex. Online dating has the same end result as every other form of matchmaking, and so our faceless they get to keep promoting real life as the primary interface, and therefore 'accurate representation' as incredibly important, and therefore 'misrepresentation' as dangerous to the whole scheme. I'd be remiss not to address the other big reason why people are so afraid of anonymity on the Internet, which is all of the times people's fears about what happens when you get to claim any identity you want have been proven rational: the children lured into malls and parks by predators, etc. I don't mean to legitimize the violence made possible by the Internet, or even to explain it, but we should be wary of how these instances get overexposed. That the potential for this violence even exists is cause for sensationalist reaction, to the extent that it is all some people know of the Internet. It's not hard to believe that a good percentage of the world population may think the Internet is a dangerous place where every unsuspecting user is eventually defrauded and/or molested. In recent years, these fears (of violence, and also of disappointment with an online dating partner) have been employed as marketing tools to sell Internet safety software and services. We must remember that although the fears of the general population have come to life, they do more good in selling products and maintaining control than they do in actually keeping people safe. To say that we must steer clear of Internet sexuality is to say that we must continue practicing the current model of sexual compartmentalization at the whim of the patriarchy. So it is no wonder that we are bombarded with examples of how not knowing who you're really talking to have led others into at least disappointing and at most deadly circumstances. Now, if they're doing such a good job of scaring people, why does anyone buy into Internet dating at all? Probably because the marketing angles are different, and yet equally effective.

What we've seen is that sexual identity on the Internet comes to encompass a few more factors than sexual identity in real life, or at least it plays these factors out in pretty different ways. Eve Sedgewick writes: 'Living in and hence coming out of the closet are never matters of the purely hermetic; the personal and political geographies to be surveyed here are instead the more imponderable and convulsive ones of the open secret.'[4] We know that there is a discrepancy between public and private identities, and that these play into performance. In public, there are two interpretations of the performance: the one we make for ourselves, and the one that is made for us. By deploying or subverting hegemonic practices, we learn to exert a degree of control over how we are perceived - how our identity is perceived, and how our position in relation to power structures is perceived - and this control is, of course, also a dimension of the performance. So, if we engage in subversive practices or identities, do we put them on display? What is an 'open secret' on the Internet? What happens to the individual who reveals some part of her identity online that she never reveals in real life? Sexuality and social networking has less to do with confession and more to do with putting yourself on a stage and not being able to see the audience. But unlike any venue for expression we've had before, you can be pretty certain that the audience is there on the Internet. You don't get to pick and choose the members of the audience, which is either frightening, exhilarating, or both, depending on who you are. Going back to Judy, she doesn't know who's looking at the MySpace profile she created, and maybe she doesn't care. What does she decide to reveal, given the faceless mass that is her audience? Is it another facet of her identity that only comes out under this very specific set of circumstances only the Internet can provide? That the Internet provides Judy with a space to act out a heretofore unrecognized dimension of identity is an important reason why we can't just draw direct correlations between her online and real life performances. Maybe it's built into Judy's identity that she presents one thing in real life and another thing on the Internet, and we now have three Judys: real life, online, and the total of both. Do we accept her online identity as the most faithful to how she sees herself, for all of the reasons I named before (that she gets to say whatever she wants, that it's total performance, etc.)? Or is her identity a combination of that which she posts on her profile and everything we saw her do getting ready in the morning? Or, again, does it depend on Judy?

Let's say, for argument's sake, that the real Judy is what she posts online. That is to say, the Internet gives us something real life never could: The ability to perform an identity that has absolutely nothing to do with biology. And if we, enlightened as we are, are prepared to abandon biology as a predicate for determining gender identity, shouldn't we be absolutely thrilled that this possibility exists? As more and more of our interactions move online, all those individuals who felt a lack of identification with their bodies get to become their true identities. We can say that the body restricts the scope and import of performance. We can draw a comparison to body-dysmorphic disorder, in which an individual perceives her body as other than it actually is. In real life, there is, perhaps for all of us, a constant sense of dysmorphia - we do not see ourselves as the politics inscribed on our bodies. And it's not even that we're inventing new bodies for ourselves on the Internet, though we can do that, too; we're expressing the identities that our bodies keep us from having in real life. And if a part of that identity happens to be that you don't need a body for sex, all the better for you. But I want to bring up Butler's question again, because it's still a problem. Earlier I quoted her saying, '…if the model of a more diffuse and antigenital sexuality serves as the singular, oppositional alternative to the hegemonic structure of sexuality, to what extent is that binary relation fated to reproduce itself endlessly? What possibilities exist for the disruption of the oppositional binary itself?' Are we just left with a new binary? Perhaps the important thing is that the Internet has brought up all of these questions that aren't going to go away. We are constantly having to ask about the discrepancy between real life and Internet identities, and why they matter. If gender theory has so far gotten us to embrace the idea that gender is a social construction, and that there's nothing wrong with feeling yourself a different gender than the one you were born with; and if queer theory has gotten us to accept any number of sexualities as valid; can new thought move us to a place where we accept Internet identity as privileged equally or even above real life performance? It's possible that Butler might say we still have the same problem

[4] Eve Sedgewick: Epistemology of the Closet. Berkeley: University of California Press 1990. p. 80.

of legible versus illegible, and simply making everything legible doesn't clear that up. She calls for the dissolution of everything politics and culture do to identity, and everything they will do on top of that. So I haven't gotten us any closer to this ultimate ideal, since just bringing Internet sexuality into the realm of the generally accepted doesn't clear up the fact that there might still be a division between accepted and un-accepted. Butler doesn't just want us to broaden the circle; she wants to tear down the boundaries. I don't know for sure that we won't, and I'm not ready to say that embracing the possibilities of the Internet is merely another step in line with a patriarchy that actually wants us to broaden the circle for whatever reason. But the difference with Internet sexuality - as opposed to what's happened with feminism and queer theory - is that it leaves absolutely nothing fixed in place. We leave open the possibility that, not only is Internet identity a legitimate presentation of the individual, but also that the individual gets as many Internet identities as she wants and that these are all legitimate. It is the tendency of hegemony to fix as many things in place as possible - even and especially if what's being fixed in place runs counter to hegemony, that is, if you're outside the circle, you stay there - and to limit people's ability to change identities. This is what supposedly keeps society in order and makes it easy to understand other people. But all the heteronormative notions about gender, sex, and identity are worn out, irrelevant, and ready to be dismantled. There will be anarchy in the sense that no old order will be able to dictate what performances and expressions are valid; there will be no conformity because there will be no guidelines.

Nathan Shedroff / Chris Noessel

MAKE IT SO (SEXY)

'Say, what's a droid like you doing in a joint like this?'

Make It So (Sexy) is a subset of larger work we're doing for a book called *Make It So*. The larger work explores how science fiction and design - and in particular *interface* design - relate to each other. Ultimately, the work explores what interaction designers can learn from the interfaces seen in science fiction movies and television shows.

Why only movies and television?

The book's focus on interfaces restricts the categories of science fiction we consider. In order to be effective, the interfaces we analyze must be visual, which excludes written descriptions in books. We must see the interfaces in use, which constrains us to moving media, which excludes graphic novels and illustrations. They must also be consistent across depictions, which exclude most hand-drawn media. These constraints leave us with television and film. Our sampling for this presentation was similarly restricted.

For *Make It So (Sexy)*, we did look at a selection of science fiction-themed pornography, but found that interfaces from movies in this domain are sorely lacking. Pornographers seem to treat science fiction as a style of costuming and an excuse to stage actors' intercourse with strange puppets. One example is *Anal Intruder 9*, in which a pair of outer-space aliens land on Earth and plot to take over the planet. Their scheme involves having anal sex with every woman they can find. Any science fiction conceit of the movie is merely set dressing, discarded when it comes to the actual act. As a result, only one moment from pornography had any relevance for our purposes.

For *Make It So (Sexy)* we looked, specifically, at sexual interfaces in science fiction and compared what we found with high-tech sex developments in the real world to investigate the relationships and possible connections. After collecting examples in science fiction, we identified three main categories for sexual technology:

- Matchmaking
- Sex with Technology
- Coupling (which further breaks down into two subcategories, i.e. Augmented and Mediated)

Matchmaking

Matchmaking technologies get people together for sex. They either let a user specify desired aspects of a mate, or help people meet other interested parties for sex. In our survey we saw the least of this kind of sex related technology in science fiction, and much of it seemed to be older.

In the film *Logan's Run* (1978), Jessica puts herself on The Circuit, a system where people interested in having sex are teleported between different residences of those seeking sex until a match is found.

In *Weird Science* (1985), Gary and Wyatt specify the physical aspects of their ideal (and fabricated) woman. They do this by inserting clippings from magazines which embody desired traits, including physical and mental ones.

In *Total Recall* (1990), the main protagonist, John, specifies the type of love interest he'd like in his manufactured memory vacation. The variables are mostly physical but touch on sexual aggressiveness as well.

In the television series *Firefly* (2005), the character Inara is a companion, a kind of highly skilled and trained courtesan whose services to her clients often include sex. To get clients, she sends notification when she will be in a location, and potential clients apply with video messages to spend time with her. The system includes collaborative filtering by the companions so that dangerous clients are excluded.

Why are there so few examples of matchmaking sex tech? We suggest that the facts of specifying a mate are very well known to the audiences through their direct experiences of dating sites. They know that one has to specify quite a bit of information to get a good match out of a matchmaking system, and that's just not very cinemagenic.

What matchmaking technology correlations do we see in the real world?

There are plenty of matchmaking sites on the Internet for sex. Adultfriendfinder.com, Manhunt.com, and Alt.com help users find matches for sex with a surprising level of detail.

iTrick is an application that lets users document incredible amounts of detail about the sex they have had with partners. It presumes a gay male user, and in the interest of thoroughness includes some neologisms in its database, such as 'Cock-suckee' and 'Jerkee.'

The Love Gety, popular for a short while in Japan, came in male and female versions. Users could set their fob for one of three settings: chat, karaoke, or 'Get2.' When a fob of the opposite gender was within 15 feet, if your modes didn't match it would blink in the 'find' mode. When they did match, it would blink in the 'get' mode.

Sites such as *TheEroticReview* let users write about their experiences with escorts in order to collaboratively filter them. In our research, we heard tell of a similar system by which escorts could review clients, but could not find reference to it. These sites may no longer be active.

In this section, we find a few correlations between what's happening in film and television and what's happening in the real world. Given some more recent developments in technology, there seems to be plenty of unexplored space in match-making technology. For example, what happens when people can collaboratively filter partners? Using ubiquitous sensors and subtle actuators, how 'magic' and subtle can proximal matchmaking become?

Sex with Technology

Sex with Technology occurs when a human has sex with technology in some form and there is no other human partner directly involved. When such on-screen technology is physical, it can range from very human like - as with sex bots that are indistinguishable from humans - to the very mechanistic (which are usually portrayed as dystopian). When on-screen technology is virtual - as in Star Trek's holodeck - we only see virtual replacements for humans.

Device-like
THX-1138 (1971) features a scene in which the male protagonist receives physical gratification from a mechanical device while watching a sexy hologram.

In *Sleeper* (1973) Luna and a guest spend six moaning seconds together in a device called the Orgasmatron.

Cyborg

In *Space Truckers* (1996) Cindy is both a little excited and terrified at the sexual prosthetic sported by the cyborg villain Macanudo.

Sexbots

There are so many examples of sexbots it is easier to simply list the shows in which they appear: Buffybot from Buffy the Vampire Slayer; Lucy Lui-bot from Futurama; Pris from Blade Runner; the 'clicker' Pax from Creation of the Humanoids; Mr. Universe's sexbot wife from Serenity; The Stepford Wives (both incarnations); The sexbot barroom whore in Westworld; Data from Star Trek: The Next Generation; Valerie 23 and Aiden's companion, each from The Outer Limits; and though not specifically created for sex, the human-like cylons in the modern reboot of Battlestar Galactica are fully capable of having sex with humans, and do so from the first episode.

Computer generated virtual sex partners

In *Star Trek: Voyager* (1999) the Doctor prescribes a tactile-holographic remedy to a Vulcan's Pon Farr despite the fact that there is no female Vulcan within light years.

Also from *Star Trek: Voyager*, Tuvok satisfies his own Pon Farr - and avoids philandering - with a holodeck version of his wife.

In *The Matrix* (1999), though we never see it, Mouse assures Neo that he can arrange a more 'intimate' experience between Neo and the woman in the red dress.

The prevalence of sexbots and holo partners is understandable, since being indistinguishable from humans, these are easy to cast, and special effects needed to underscore the non-humanness can be minimal. Other technology with which people have sex are much rarer and almost always have a clear bias about the act (the world in which the sex occurs is almost always shown as dystopian, or the individual having the sex with tech is seen as incompetent or immature).

What sex with tech correlations do we see in the real world?

Real world examples in this category pale in comparison to the examples on screen.
The closest we can get to the physicality of sexbots is the Real Doll (which seem to the authors deep in Mori's 'uncanny valley.')[1]
The current state-of-the-art in Robotics barely approaches truly human-like behavior. In March of 2009, Kenji, a caregiving robot created by Toshiba's Akimu Robotic Research Institute, had to be disabled after it trapped a research assistant trying to give her hugs.
Non-human tech devices abound and are taken much more seriously than what is seen on screen. OhMiBod is a music-powered vibrator that plugs into an MP3 player that moves to the rhythm and intensity of the music.

---

[1] The 'Uncanny Valley' is a theory by created by Masahiro Mori to describe the revulsion many people often feel for human facsimiles. As the humanlike representation improves, generally, people are more attracted to devices. However, as the accuracy also improves, at once point, revulsion sets-in and these facsimiles are no longer accepted because they are too accurate (without being animated). When the representation matches a real human, however, our revulsion often dissipates.

The BodiTalk (by the same company who makes the OhMiBod) is a similar device that plugs into a mobile phone, and is triggered to vibrate to a pattern when the phone rings.

Other non-human sex tech devices include the devices featured in fuckingmachines.com, the Fleshlight, and the remarkable prototype RealTouch, which is like a Fleshlight with actuators inside for warmth, lubrication, and motion each of which is synchronized to specially-coded online videos.

The closest we come to virtual sex partners are with branded websites, where for example, a user can make his or her avatar have sex with an avatar of Jenna Jameson.

Coupling

We found two subcategories of coupling technologies. The first - and there was only one example of it - is augmented coupling, where technology enhances an otherwise analog physical act. Interestingly, this is the one example that came from pornography.

Augmented Coupling

In *SexWorld* (1978), a pornographic send-up of the then popular *WestWorld*, a client is astonished when sexy music begins to play automatically in the bedroom in which he is being seduced.

The second subcategory is *mediated* coupling, where two partners have sex with technology as the enabling media, often precluding any physical contact. This subcategory showed the most interesting and forward-looking examples.

Mediated Coupling

In *Barbarella* (1968) Barbarella and Dildano have sex 'like they do on Earth' by taking a pill and touching hands. Though they remain physically very still, they have deeply moving internal experiences.

Also from *Barbarella*, Barbarella undoes Duran Duran's murderous plans by outlasting his sexual torture device, called The Machine.

In *Flash Gordon* (1980), Ming uses his ring to enthrall Dale, using a gestural control to test her sensual response as a prerequisite for marriage.

Demolition Man (1993) shows John being surprised when Lenina's frank offer for sex turns out to be a virtual one, accomplished through two paired helmets.

Lawnmower Man (1992) features a scene in which Jobe treats Marnie to a psychedelic introduction to cybersex.

In the television series *The Outer Limits*, episode 'Skin Deep' (1995), Sid wears a handsome holographic disguise, with which he seduces his love interest.

Brainstorm (1983) follows the development of a sensory recording device. In one scene Michael rescues Alex from a stupor induced by a looped full-sensory sex recording.

In *Strange Days* (1995), Lenny is a peddler for amateur full sensory recordings that most often include sex.

What coupling technologies correlations do we see in the real world?

Interestingly, the real world examples of coupling, while interesting, are only cursorily related to what's being shown on screen. In other words, there seems to be a noticeable divergence between the fantasies portrayed on screen and the devices actually being built for use.

There are virtual costume and transformation fetishes, as Malebots.com and detailed virtual Furries illustrate.

Virtually, *Second Life* and (the more built-for-purpose) *Red Light Center* allows users to have sex with each other via avatars with screen controls (and, often an audio channel).

The *Thrillhammer EC01* is the world's first commercial teledildonic sex machine. It allows Web surfers to watch video and remotely control actuators that pleasure the performer reclined in the device.

Fucking machines, such as those shown at *Arse Elektronika*, embody a very physical mediated coupling.

Greg Larson's *The Machine* lets one person pedal a chain that drives a dildo in and out of another person sitting on the nearby seat.

Benjamin Cowden's *Eating My Cake and Having it Too* offers a user to turn a crank that causes a self-lubricated tongue to gently lick a lollipop mounted in place. One presumes that the device could have non-lollipop applications as well.

CoxyPro's *Ultimate Sexual Arousal Vehicle* is a marvel of video and physical mediation between the operator and the recipient.

Given the amazing variety and ingenuity we see in the devices being built in reality, there seem to be many unexplored opportunities in screen-based science fiction. One that comes quickly to mind is the number of participants. With network technologies, why is coupling limited to 'one to one' experiences? (Even the *Thrillhammer EC01* was a turn-taking device.) What happens when 'many' can participate at once, either as controller, experiencer, or in both roles?

Nonetheless, though there certainly are many impressive technologies for coupling in cinema and the real world, they seem to have little to do with each other. Neither seems to be leading the other as we have seen in other technology/science fiction domains such as security and telecommunications.

The larger cycles of influence

In the larger work, Make It So, we describe the process in which products and services in the real world establish a technology paradigm, which science fiction extends in limited ways. In this way, current technologies establish the context for and ground our understanding of future visions.
In turn, science fiction influences the development of real interfaces in four primary ways.

- Inspiring individual engineers and designers
- Setting expectations in the audiences
- Reminding the audience of the human context
- Proposing wholly new paradigms

Each of these influences is explored more fully in *Make It So*, but this summary helps us understand a little more of what we've seen in science fiction sex technology.

Looking at sextech, there's not enough parallelism between the screen and real life to assert that any but individual influences are taking place. This is understandable since film and television studios have to walk a fine line between titillation while, at the same time, keeping a low and broad rating, since they're mostly looking for large audiences and economic success. American audiences set the standards of the ratings, and their general prudishness and sex-negativity constrain the studios from pushing too far.

'Sexplorers' on the other hand, are genuinely interested in pushing the boundaries of sexuality using technology; playing with gender, species, the effects of the network, and even crossing senses with synaesthesia. Truly, these hobbyists are leading the way. But, why is this?

One reason is that sexual technology often 'breaks' science fiction stories (in films and television, at least). Most sex-related interfaces require explaining, taking away from the plot and reducing any titillation. (And given the prudishness mentioned above, titillation is the most studios are willing to risk anyway.)

But, there may be something more fundamental at work here. The sex imperative is millions and millions of years old, and deeply coded. The sex drive sits at the kernel, not (just) the presentation layer (to abuse a computer metaphor). The human race wouldn't have made it as far as we have if this drive weren't strong and mostly tamper-proof. Most of sex resists the intrusion of anything not-sex, and for most people, that includes technology. This may explain why pornographers such as those responsible for *Anal Intruders 9* simply discard technology in their films, too.

If this is correct, it will be up to the hobbyists (and authors, though our survey doesn't currently include them) to continue to lead the way in extending and exploring new roles of technology in sex.

Sharing is Sexy (DJ Lotus / Kelly Lovemonster / J Bird / Scrufty Eudora)

RADICAL PORN
INTERCOURSE BETWEEN FANTASY AND REALITY

Kelly Lovemonster: Hello.

DJ Lotus: Hi.

Kelly Lovemonster: I am Kelly Lovemonster, and I am with *Sharing is Sexy*. Today we are going to talk about things.

DJ Lotus: Hi, I'm DJ Lotus and I'm also with *Sharing is Sexy*. We don't do lots of these types of talks. We do other things, so if this is a little clumsy that's why.

Kelly Lovemonster: Who is *Sharing is Sexy*?

DJ Lotus: We are a radical porn collective, and we make porn. That's me on the right.

Kelly Lovemonster: And that's me over here.

DJ Lotus: But *Sharing is Sexy* is much bigger than us. There is about six or maybe eight of us.

Kelly Lovemonster: Yeah, *Sharing is Sexy* also puts on events throughout San Diego City. We host porn watching events, we put on burlesque shows…what else do we do?

DJ Lotus: We make zines.

Kelly Lovemonster: Yeah, we make zines. Overall, we make radical porn.

DJ Lotus: We also called this project our open source laboratory because all the work we do is under the creative common's license, so it's all free. You can go to SharingisSexy.org to see all the wonderful sexy photos and videos. And we try to use open source methodologies and apply them to porn production. And we will talk about that a little more as we continue.

Kelly Lovemonster: Okay, to change things up a little bit we are going to do a breathing exercise with everyone. It is really easy. It's three parts; it's going to require you to get up out of your seats and stand up. So, if everyone could do that for me. Thank you. So, I'm going to explain how this is going to work and then we are going to do it. First, we are going to close our eyes, and we are going to take four deep breaths in. I want everyone to really focus on their exhalation pushing all the air out of their lungs. And once we are done with that, we are all going to start silently chanting 'no' together. Similar to this 'no, no, no, no, no…' And I really want everyone to pay attention to how 'no' feels in your body. And we are going to progressively get louder and louder with our 'no(s)' until we are all shouting 'no.' Once we are all yelling 'no,' we are going to switch over and begin shouting 'yes.' And I want everyone to pay attention to how the 'yes' feels in your body. And I want everyone to really experience their 'yes.' If your 'yes' takes you to a place where you have to touch yourself, or if your 'yes' takes you to a place where you need to cry, or if your 'yes' takes you to a place where you need to scream, I want everyone to experience that. Okay?

DJ Lotus: Who's going to switch to 'yes?'

Kelly Lovemonster: We will (laughs.) So, here we go. We are going to start with our deep breathes in and out. Everyone: deep breath in…and I want you to push that deep breath out. Another deep breath in… and I would like you to push that deep breath out. Deep breath in… and a big deep breath out. And last deep breath in… and push all the air out.

Everyone in unison: (Progressively getting louder and louder) No, No, No, No, No …No! No! NO! (Shouting) Yes! Yes! Yes! Yes…Yes! Yes! Yes!

Kelly Lovemonster: That feels good. Thank you everyone. (Audience Applause)

DJ Lotus: So, now we are going to talk about what's real. Perhaps that was a little hint for everyone.

Kelly Lovemonster: We are going to setup some parameters for our talk. We've defined some words; so, we can all be on the same page with how we are going to use language. This way we can all have the same conversation. Anything else you would like to include with that.

DJ Lotus: Sure, we have lots of slides, some photos, and a few short video clips. We are going to look at other projects and ideas about the real, how it's used, and what it means.

Kelly Lovemonster: Defining the real. What does the real mean to you? I want to know, when I use the word real what does everyone think about the real. Anybody? Yell it out. What does the real mean?

Audience Participant 1: Not Fake!

Kelly Lovemonster: Not Fake. Anyone else?

Audience Participant 2: Everything!

Kelly Lovemonster: Everything is real. I like that. Anyone else?

Audience Participant 3: Orange Juice!

Kelly Lovemonster: Orange Juice is very real. Anyone else?

Audience Participant 4: Now and instinct!

Kelly Lovemonster: Those are very good. Anyone else. Anything else about the real?

Audience Participant 5: Consciousness!

K and DJ Lotus: Consciousness!

Kelly Lovemonster: Consciousness is real.

DJ Lotus: Anything from the tech booth? What's real?

Tech Booth: Technology!

K and DJ Lotus: Technology!

Kelly Lovemonster: Technology is real.

Audience Participant 6: Shit!

Kelly Lovemonster: Shit is very real.

DJ Lotus: Okay, lots of different ideas about what is real and what it means. We thought we would just gather some accepted and common notions of the real.

Kelly Lovemonster: Phen- om- o- logical? That's a hard word.

DJ Lotus: I would say this is the approach we are mostly using for this talk, a phenomological approach to the real. What your feelings and sensory perceptions are, we would say those are real. That exercise we did, we can say that is something that is real.

Kelly Lovemonster: Here is a random house unabridged Webster's dictionary definition of the real. Genuine, not counterfeit, artificial, or imitation, authentic.

DJ Lotus: Someone said something like that over there. Not fake. But again I really appreciate the previous talk, but we were thinking 'that's our talk, shit!' So here, we have a little complication of what is real and what is not from Lacan and psychoanalysis. Lacan has these three levels: the imaginary, the symbolic, and the real. And his idea of the real is that the real is inaccessible to us, that raw physical materiality we can never get to. Just keep dreaming. Every experience even the most intimate sexual experience is filtered through your idea of who your touching, what your touching, and your being touched, your perceptions of what does pleasure mean.

Kelly Lovemonster: Thus being not real?

DJ Lotus: So, now we are going to go through a few different porn projects and look at the way they use this idea of real. Part of the motivation for this talk was to add to this conversation of what is radical porn by addressing this one point of realness.

Kelly Lovemonster: We would like to put these porn projects side by side and ask the question of if the real can exist in all these projects simultaneously.

DJ Lotus: Continuing on with a few more definitions. And in trying to make this quick…radical, what the hell do we mean by that? We are definitely taking the easy way out with that for now. And we are relying on what projects what these projects say about themselves.

Kelly Lovemonster: All the following words come from all the projects we are going to go through. Some projects claim to be grass-roots, feminist, sex-worker friendly. You guys can read the rest. Point being this is the language they use.

DJ Lotus: Now, a quick definition of porn. A definition that we think is rather crappy from the United States Code, Title 18, Section 2257 which is the fine print of the bottom of every porn site you look at, the complicated laws for documentation. And so that one is: 'a visual depiction of actual sexually explicit content', I don't know what that means. But it is a good example of what is real, or what is actual. It is right there in the legal definition of what is pornography and what is not.

Kelly Lovemonster: This is what SiS thinks porn is: anything that one thinks is sexy, and wants to share. So, if I want to take off my shirt right now and touch my nipples, that's porn, it's sexy to me. (Laughs)

DJ Lotus: We'll wait. Let's hold on that.

Kelly Lovemonster: OK.

DJ Lotus: So now we're going to talk about a few specific projects that elicit or use this idea of the real. So for this section we're talking about a few different projects that sort of rely on the real, that try to be sexy or be radical by saying that they're something that you can relate to.

Kelly Lovemonster: Yeah, 'these people are like you'.

DJ Lotus: Our first example that we'll talk about a bit is Annie Sprinkle. Annie Sprinkle is a porn performer with a great video, *Herstory of Porn*, and we'll show a bit of that. The video goes through all the different porn projects she has worked on and the first one we'll show is from the early 1980's. It was an attempt to make porn radical by having it be made by women and be more psychological.

[Clip from *Deep Inside Annie Sprinkle*]

> Annie Sprinkle - 'I wanted to make a movie where the people sitting watching it weren't just sitting watching, something where they were a little more intimately involved, something kind of interactive, so throughout the film I talk directly to the camera.

> 'Hi, I'm Annie Sprinkle. How are you? I'm really, really glad that you came to see me because well I want to get to know you and I want you to get to know me and I want us to be very, very close, and very, very intimate. Would you like that? I would.

> 'You know, I am a real exhibitionist and I love to have sex in very public places. Sometimes, if I see one of my movies is playing, I'll go in and sit down and start watching myself fucking and sucking on that big screen, and that makes me very, very horny. So, I get carried away and I start doing all the guys around me. It's really nice. So don't be surprised if you're ever in a movie theater and I come in and I sit right next to you.

'So you see, now it was I that became the sexually aggressive one. No one had to force or manipulate me into sex. I was the one who wanted it and the guys, they better just watch out.'

DJ Lotus: OK, as much as we'd love to watch the rest of this video, it proceeds just as she said; she goes in and has sex with lots of dudes in the movie theater.

Kelly Lovemonster: See, that could be you! She could walk in right now, and sit right there, and then sit on your face.

DJ Lotus: Moving on, then we have Suicide Girls. This gets to the question of 'what is radical porn?' Part of the parameters of this talk for me is trying to avoid binary distinctions, so I'm not going to try to say 'this is radical, this is not, this is porn, this is not.' But, Suicide Girls calls themselves grassroots.

Kelly Lovemonster: There's a quote there by the New York Times that we thought was appropriate and we wanted to highlight, 'the meeting place for people interested in alternative lifestyles.' So, you can go on Suicide Girls and you'll meet girls who are alternative and who are like you. Again, this is not to say that you wouldn't, but...

DJ Lotus: That's the claim right, the meeting place for people, so you're just like them. You might run into them in the street. You're cool and alternative, so are they. They might fuck you.

Kelly Lovemonster: Then, NoFauxxx.com. We like No Fauxxx. I think that just the title of their website begins to imply that these people are not fake, they're real. They are like you. You too can submit your photo and be on No Fauxxx and show your sexy porn. I'm still waiting to hear back from No Fauxxx. (Laughs)

DJ Lotus: Not only do we like them, I would say that No Fauxxx is the site that is most like our site, it has all different sexualities represented and genders and they're not categorized by gender. The difference is that you have to pay for No Fauxxx's site. Obviously they're big on the 'not fake' thing.

Kelly Lovemonster: And, *Enough Man*.

DJ Lotus: So, another short clip, dealing with the real, another porn project.

Kelly Lovemonster: *Enough Man*, this is a documentary about F to M (Female to Male transgender) people and it shows them having sex and talking about their sex and how they experience their sex and sexuality.

DJ Lotus: The back of it says 'documentary meets porn' and it's basically that format, shot in documentary style, as you will see.

Kelly Lovemonster: This clip is a little graphic, so ahh, that's all I have to say.

DJ Lotus: And maybe brings a different idea of real.

Kelly Lovemonster: Yeah definitely. I think this is really real. (Laughter)

[Clip from *Enough Man*]

> 'I think that for a lot of transgender folks that's really important, a lot of transgender/transsexual folks, because it's not done and it's not seen and so many people have not been able to be sort of up close and personal and see how other transgender people fuck, you know?

> 'With needle play there's not a lot of blood and it's not really super invasive.'

Kelly Lovemonster: So that was some real F to M porn right there.
Ok, so now comes, *Sharing is Sexy*, us, how we elicit the real. The idea of our site is that you can come on over and shoot some porn, after you hand in your proper documentation and you too can be on the site. *Sharing is Sexy* is just real people who want to put on the internet what they think is sexy.

DJ Lotus: I would say there's a sort of built-in assumption about realness in the fact that it's DIY (Do It Yourself), Open Source. All those things are implied, that it's open, I mean, it is. You can submit porn to us. We can talk about it after the talk. But regardless, we are also relying on a similar claim.

Kelly Lovemonster: Ok, so, Fantasy defines the real but it doesn't undefine the real.

DJ Lotus: For this next section we'll talk about some projects that are more fantastical, that are not trying to be so directly real like these projects were.

Kelly Lovemonster: We're going to propose or ask, what does it mean for our fantasies to turn us on in the real? What does it mean to have these fictional things elicit real feelings within us?

DJ Lotus: And we'll talk more about how fantasy operates in these projects and brings about arousal. Ok, here we have some more quotes from the old dead European guy. For Lacan... just to bring this in and continue to complicate this discussion of the real, for Lacan, fantasy shapes the real. We have fantasies that shape what our notions of pleasure are. We have these abstract ideas, basically, that we use to interpret our interaction with the materiality of the world.

Kelly Lovemonster: We use a phenomenological approach to porn. We're not trying to say we have all the answers. Again, what gets us off might not get you off. What we find sexy you might not find as sexy, and that's okay.

DJ Lotus: Here's a nice little sci-fi-ish clip that we'll show from Mandy Morbid, who has a claim of being radical by being anti-capitalist. The banner on the top of her website used to be 'All of the Pussy and None of the Capitalism,' which I can get behind, I dig.
Again, like the rest of our stuff, this clip is graphic too. But, whatever, you paid for this conference. The title is Mandy Morbid vs. Crazy Tentacle Sex Monster.

[Clip from *Mandy Morbid vs. Crazy Tentacle Sex Monster*]

DJ Lotus: Ok, ok, as much as we would love to watch the rest of this video, we'll have to interrupt.

Kelly Lovemonster: So again, that obviously wasn't real, right? But still, there's some thing that was really real about that. I mean, that was sexy, that was hot. There are tons of kids across the world masturbating to Mandy Morbid being fucked by a weird tentacle object.

DJ Lotus: And here we go back to the weird picture of Hegel, and repeat 'what gets us off might not get you off.' My experience of that video is that it's pretty hot.

Kelly Lovemonster: On to *Lickety Split*.

DJ Lotus: Ok, *Lickety Split* is another radical porn project that is based on Montreal and they put out a zine. They call it a pansexual smut zine, so we're not going to show you any exciting video.

Kelly Lovemonster: We thought it was really important to include this because it is made up of words. It's the idea of fiction through language getting people off. For those people who love to read their porn. I remember some of my first experiences with porn were with porn through the internet, but porn stories. I thought 'oh, so hot!' and I would jack off to all these weird erotic stories. That to me was very real.

DJ Lotus: Surely, there's a long history of erotic fiction and we could say that it's porn and an example of fantasy getting people off.
So we're going to show one more clip at the moment from Pink and White productions. This movie is called *Superfreak*. Pink and White is, what does the top of their website say, 'porn for pussies'. It's lesbian porn made by women, presumable for women, made here in San Francisco.
This video is another example of something that's clearly not real. The main plot line of Superfreak is that Rick James' spirit inhabits people's bodies and then they have really wild, amazing sex. So, we'll show a little bit of that. But even though we know that Rick James' spirit isn't really inhabiting their bodies, somehow we are still aroused, possibly. Oh yeah, we forgot to say the expert thing...

Kelly Lovemonster: Oh yeah, we're not experts.

DJ Lotus: We're not claiming to be experts, we're all about DIY. We're only experts in the sense that everyone in this room is an expert. Maybe we're experts at what gets us off. Hopefully you are too. I mean, at what gets you off, not me.

[Clip from *Superfreak*]

'Ah, that's freaky!'

DJ Lotus: Another good video.

[Slide of Image from Jen Vilotta]

Oh, that was too early! So here we are talking about fantasies and porn that is about fantasy and maybe something good about porn being that people get to fulfill their fantasies, or that people can learn to fulfill their fantasies if we think about porn as pedagogical. If we think about some person living in Omaha who's never met a transgender person, they might see some transgender porn and think 'hey, that's me!' In that way we could see fantasy being good.
But we just want to complicate the ethical discussion by admitting and bringing up this fact.

Kelly Lovemonster: So here we have some porn made in *Second Life* by Jen Vilotta. So, is this right? Is this sexy? I don't know. Is it real?

DJ Lotus: Clearly it's not real. She puts all her stuff on *Flickr.com*, so if you go to *Flickr* and look up her name you can come up with her stuff and there are tons of comments form people about how wonderful it is, how erotic it is, how sexy it is. It's not all blown off heads, but it is lots and lots of violence and severed arms and such things.

Kelly Lovemonster: I think we thought this was important because it really does ask that question 'what porn is okay? What porn is okay to show?'

DJ Lotus: It also brings up another interesting thing about this image is that *Second Life* is a perfectly good example of really cheap production, (30:12) right? Here's a platform that people have where they can make whatever fantasy they want come true. They want to be furries, dragons whatever they want to be, they can do that, and in a lot of cases, there's work like this.
In this dialog about what's real, we also don't want to pretend like it is our brilliant idea or that we brought it up for the first time. There are a couple of porn projects that play with the idea of the real or focus on the fact that they're not 'really real' or that 'real' is, maybe, silly.
For example, *In Search of the Wild Kingdom* is a faux-documentary that is about a documentary crew who are trying to find the 'real' sex lives of 'real' lesbians in San Francisco. The camera crew and the director and prominently featured in the film. The lights and the cameras are often times in the shot. They are very consciously making a joke about worrying about what's real.

Kelly Lovemonster: This is another documentary, where in the same vein; they show you the production of a porn film. The *East Vancouver Porn Collective* is a small collective and they make porn and they make fictional porn. We don't know if this movie is a documentary, or a porn film about making a documentary, but they definitely bring up the question.

DJ Lotus: I think they're a good example of something that Annie Sprinkle does well where you're just not really sure. Their movie, *Made in Secret,* is about them making porn, but it doesn't actually feature any sex in the movie and they never released the movies. They never have actually admitted whether or not they really made the movies. They're really pushing the real-ness question. There's some making out in the movie, but there's no, say, vagina.
We're going to show a little bit more of Annie Sprinkle because we love her and she's a big inspiration to us. I think this is another example of porn which plays with the real, which is on the borderline of what's real and what's not.

Her video *Herstory of Porn* goes through all the different kinds of porn that she's made. Starting in the seventies, where she was talking about people exploiting her. Then moving on to all kind of stuff like S&M porn, tantric porn and then getting to this.

At the end of this movie, *Herstory of Porn*, Annie Sprinkle goes through a little bit of DIY porn, which is sort of what we do. She tries to go step-by-step and tell you how to make 'porn made easy'. You get a director. You get a star. She says 'call up a friend who you want to have sex with.'

In this particular scene, she shows some of the production methods.

[Clip from Annie Sprinkle's *Herstory of Porn*]

Annie Sprinkle – 'I would want the camera to move in tight on all of the erogenous zones.

'Bubbles! I want lots of bubbles!

'I could get my friend Melanie to do an erotic fish dance.

'This is Serena. She learned about me in her women's studies course. She wanted to apprentice with me, so I put her in my porno movie.

'Make sure everyone is over 18 years old and be sure and get a signed model release'.

(38:49)

DJ Lotus: Here we can say that there are many different levels of the real and fantasy interplaying.

Kelly Lovemonster: Yeah, that penis...

DJ Lotus: Yeah, so you might not even know that it's not real. I don't think that they showed it there, but later they even show that it's a squirting dildo [that they're using to make this scene], that has a little cumshot action.

Oh, and clearly, they're not mermaids.

For a long time radical porn has said 'we're radical because we use real people! No boob jobs and fake nails here.' Or, 'we're real because there's real female pleasure, real orgasms.' But maybe there's a problem with that, what might be some problems with this?

And then there's Lacan who says that we don't have access to the real in the first place, so if we're saying that what makes these projects radical is that they're real, then maybe we're just fooling ourselves.

Kelly Lovemonster: Take the whole idea of the accommodation of sex for production. If you've ever made porn, then you know that it takes work! To get a good shot, to make it look sexy, to genuinely make it look hot. I don't know about you guys but when I have sex, sometimes it doesn't look as hot as it looks on the website. It comes down to me having to hold me leg up here, and hold my head back-

DJ Lotus: hold it! hold it... hold it...

Kelly Lovemonster: -and to hold that face until...

DJ Lotus: hold it!

Kelly Lovemonster: -and then we get the shot. There are all these things that we have do to make our porn look 'real' and sexy and hot.

DJ Lotus: Like, 'can you just change the angle of the dildo? I can't really see it from where it is.' There's lots of that.

Kelly Lovemonster: So this is a film, [*Xana and Dax* from *Comstock Films*] and it really highlights it. It says 'Real people! Real life! Real Sex!' So the idea is that because these people are in real relationships; they know each other so their sex is more real than the sex that you would see in mainstream porn.

DJ Lotus: Comstock has a series of films that is claims to show 'Real people having sex the way they /really/ do', and it's theoretically hotter. That's cool that they're doing that. But in a way it also privileges heteronormative, monogamous couples. So the sex that I have with someone that I've been with a long time is more real than the sex I might have with somebody at Folsom [Street Fair]? I don't know.
We're talking about what's real and that being a basis for radical porn. Who's excluded from this? Who gets to make radical porn?

Kelly Lovemonster: She doesn't get to make radical porn.

DJ Lotus: Clearly, Number 6 [from the TV series *Battlestar Galactica*] doesn't get to make radical porn, or anybody else who's not a 'real girl'. Suicide Girls in their model application form says that they only want real biological girls. What does that mean? No cyborgs.
We can see with the examples that we've looked at that lots of radical porn has lots of body modification in it.

Kelly Lovemonster: Definitely, there are tons of people with tattoos and piercings and there's an assumption that these types of body modifications are okay and these types of body modifications are very real, as opposed to, I don't know, breast implants. We might watch some women in porn with boob jobs and all of the sudden that porn becomes less real to us.

DJ Lotus: Nikki Sullivan has a really good article called 'Transmogrification' that has a nice detailed analysis of the rhetoric which says that boob jobs are fake but tattoos are real, or boob jobs are a product of people succumbing to patriarchy while tattoos mean, somehow, that people have agency.
With this issue of boob jobs, we mean cosmetic surgery in general. So on a more serious level of asking who's excluded, if we're going to say that people with boob jobs are not real and can't make radical porn, then I would say that that excludes a lot of sex workers. A lot of sex workers do get cosmetic surgery because they make more money that way. It makes their lives a little easier.

Kelly Lovemonster: And, we brought up transgender people, some of who go through surgery and get boob jobs or lip collagen and lots of other things. So that makes them less real and they sex that they have less real?

DJ Lotus: I'll bring up the caveat that you asked me to bring up, which is… ostensibly there are no transgender women on Suicide Girls. But, of course, lots of transwomen are stealth and we wouldn't really know, necessarily. But, according to their little model form, if you're a transwoman, don't even bother applying.

Tattoos, piercings, bleached anuses is another thing that people in the porn industry do.

Another reason why we would say this claim that radical porn is real is problematic is that: it is just a photo.

Kelly Lovemonster: It's just a representation. We're just trying to show what we think is sexy.

DJ Lotus: You're not actually having sex with me when you're looking at those photos of me, right? Or the films, you didn't actually have sex with Annie Sprinkle or Mandy Morbid today. Technology is not there yet.

Kelly Lovemonster: I love this photo. Here's another example. This is J and she makes porn for *Sharing is Sexy* and you can go on the site and see her. It might look like she's really enjoying this orgasm that she's having. But little do you know… this shot took a lot of work! It really did.

DJ Lotus: I wanted to bring this up because I think this set is hot. People have told me that they think it's hot. In reality, when we were making this, it was sort of like, 'so what should I do next? I don't know, put your hand over there. Okay. Does that look good? Yeah. Maybe I should move my leg over here.'

Kelly Lovemonster: So what makes radical porn radical? Sorry to disappoint, but we don't have the answers. We don't know.

DJ Lotus: It's a huge discussion. We're just making one little interjection. But, if we're going to say that maybe we should stop using what's real as a measure of what's radical, maybe we should stop saying who's real and who's not, then what else are we going to talk about? Some suggestions…

Kelly Lovemonster: This is how we make radical porn or how we think we make radical porn. We have consensus. We have this collective decision making process, which can be long and tedious, but it, to us, is a radical approach to making porn. All the performers have control over all of their sets.

DJ Lotus: This is our little intervention into the discussion of radical porn, that maybe instead of talking about the real, we should talk about how it's actually made and who has a say over what happens. We propose that other porn projects could operate like we do, where the performers have final say over everything.

Kelly Lovemonster: There's a lot of talk about the question of if you charge people for your porn, is it still radical? Part of the conversation that *Sharing is Sexy* had is that it was really important to us for our porn to be free, so that it was accessible to people. Accessibility was radical to us.

DJ Lotus: I think in the early days we started talking about the project saying that it was an anti-capitalist project, and then we talked about it for hours and hours and hours and we thought, 'maybe not?' I mean, we buy lube, dildos. We pay for hosting. We say to people that there's no money exchanged, and there's not. I don't pay Kelly, nobody pays us. But, there's still money exchanged.

Another thing that we do and we propose would make porn radical is the question of licensing and access. The porn that we make, we try to make accessible to anybody who wants it, by making it Creative Commons licensed and making it free.

Kelly Lovemonster: Again the idea of low cost, accessibility, anyone can make DIY porn, right?

DJ Lotus: Although we do explicitly say on our website and are saying now that we're not saying that all porn should be free. We don't want to discredit sex workers or people who get paid to make porn. I think it's fabulous that people get paid to make porn, but we're trying to make some proposals about what approaches might work.

Kelly Lovemonster: Definitely. I think that another radical approach that *Sharing is Sexy* uses in our porn making process is that a lot of the people who make porn with *Sharing is Sexy* identify as queer or genderqueer and they share a mission of spreading queerness. We have particular goals. How would you articulate that?

DJ Lotus: I think this is another complicated one too, which is why it's third, because surely with Katy Perry and Tila Tequila and *America's Next Top Model* having a transgender woman contestant, maybe just being queer is not all that radical. But, I do think that looking at porn pedagogically, getting people to think about new possibilities is a radical thing.

Kelly Lovemonster: Other approaches that the porn projects we talked about before use to highlight that they're radical include a feminist approach. The idea that women are in control of the porn that they make, or just people having power, looking at the idea of power dynamics and just being in control of your content.

DJ Lotus: These are some other claims or ideas that are maybe a little bit more nebulous, that are not as specific, so harder to make into an approach, but have plenty to talk about.
Consent is a good one. I think that *Lickety Split* is the only project I know that really focuses on consent and educating people about consent. In most mainstream porn, you would assume that sex happens when you walk into the kitchen, take off your clothes and start fucking and there's no need for conversation. You wouldn't say in a mainstream movie, 'can I touch you?' You would just do it. That's not so cool.

Kelly Lovemonster: And that's the end of our talk.

DJ Lotus: We want to say thank you to everyone in *Sharing is Sexy* and to *Arse Elektronika* for inviting us.

James Tiptree, Jr.

AND I AWOKE AND FOUND ME HERE ON THE COLD HILL'S SIDE

He was standing absolutely still by a service port, staring out at the belly of the *Orion* docking above us. He had on a gray uniform and his rusty hair was cut short. I took him for a station engineer.

That was bad for me. Newsmen strictly don't belong in the bowels of Big Junction. But in my first twenty hours I hadn't found any place to get a shot of an alien ship.

I turned my holocam to show its big World Media insigne and started my bit about What It Meant to the People Back Home who were paying for it all.

'– it may be routine work to you, sir, but we owe it to them to share –'

His face came around slow and tight, and his gaze passed over me from a peculiar distance.

'The wonders, the drama,' he repeated dispassionately. His eyes focused on me. 'You consummated fool.'

'Could you tell me what races are coming in, sir? If I could even get a view –'

He waved me to the port. Greedily I angled my lenses up at the long blue hull blocking out the starfield. Beyond her I could see the bulge of a black and gold ship.

'That's a Foramen,' he said. 'There's a freighter from Belye on the other side, you'd call it Arcturus. Not much traffic right now.'

'You're the first person who's said two sentences to me since I've been here, sir. What are those colorful little craft?'

'Procya,' he shrugged. 'They're always around. Like us.'

I squashed my face on the vitrite, peering. The walls clanked. Somewhere overhead aliens were off-loading into their private sector of Big Junction. The man glanced at his wrist.

'Are you waiting to go out, sir?'

His grunt could have meant anything.

'Where are you from on Earth?' he asked me in his hard tone.

I started to tell him and suddenly saw that he had forgotten my existence. His eyes were on nowhere, and his head was slowly bowing forward onto the port frame.

'Go home,' he said thickly. I caught a strong smell of tallow.

'Hey, sir!' I grabbed his arm; he was in rigid tremor. 'Steady, man.'

'I'm waiting . . . waiting for my wife. My loving wife.' He gave a short ugly laugh. 'Where are you from?'

I told him again.

'Go home,' he mumbled. 'Go home and make babies. While you still can.'

One of the early GR casualties, I thought.

'Is that all you know?' His voice rose stridently. 'Fools. Dressing in their styles. Gnivo suits, Aoleelee music. Oh, I see your newscasts,' he sneered. 'Nixi parties. A year's salary for a floater. Gamma radiation? Go home, read history. *Ballpoint pens and bicycles –*'

He started a slow slide downward in the half gee. My only informant. We struggled confusedly; he wouldn't take one of my sobertabs but I finally got him along the service corridor to a bench in an empty loading bay. He fumbled out a little vacuum cartridge. As I was helping him unscrew it, a figure in starched whites put his head in the bay.

'I can be of assistance, yes?' His eyes popped, his face was covered with brindled fur. An alien, a Procya! I started to thank him but the red-haired man cut me off.

'Get lost. Out.'

The creature withdrew, its big eyes moist. The man stuck his pinky in the cartridge and then put it up his nose, gasping deep in his diaphragm. He looked toward his wrist.

'What time is it?'

I told him.

'News,' he said. 'A message for the eager, hopeful human race. A word about those lovely, lovable aliens we all love so much.' He looked at me. 'Shocked, aren't you, newsboy?'

I had him figured now. A xenophobe. Aliens plot to take over Earth.

'Ah, Christ, they couldn't care less.' He took another deep gasp, shuddered and straightened. 'The hell with generalities. What time d'you say it was? All right, I'll tell you how I learned it. The hard way. While we wait for my loving wife. You can bring that little recorder out of your sleeve, too. Play it over to yourself some time . . . when it's too late.' He chuckled. His tone had become chatty - an educated voice. 'You ever hear of supernormal stimuli?'

'No,' I said. 'Wait a minute. White sugar?'

'Near enough. Y'know Little Junction Bar in D.C.? No, you're an Aussie, you said. Well, I'm from Burned Barn, Nebraska.' He took a breath, consulting some vast disarray of the soul.

'I accidentally drifted into Little Junction Bar when I was eighteen. No. Correct that. You don't go into Little Junction by accident, any more than you first shoot skag by accident.

'You go into Little Junction because you've been craving it; dreaming about it, feeding on every hint and clue about it, back there in Burned Barn, since before you had hair in your pants. Whether you know it or not. Once you're out of Burned Barn, you can no more help going into Little Junction than a sea-worm can help rising to the moon.

'I had a brand-new liquor I.D. in my pocket. It was early; there was an empty spot beside some humans at the bar. Little Junction isn't an embassy bar, y'know. I found out later where the high-caste aliens go – when they go out. The New Rive, the Curtain by the Georgetown Marina.

'And they go by themselves. Oh, once in a while they do the cultural exchange bit with a few frosty couples of other aliens and some stuffed humans. Galactic Amity with a ten-foot pole.

'Little Junction was the place where the lower orders went, the clerks and drivers out for kicks. Including, my friend, the perverts. The ones who can take humans. Into their beds, that is.'

He chuckled and sniffed his finger again, not looking at me.

'Ah, yes. Little Junction is Galactic Amity night, every night. I ordered . . . what? A margarita. I didn't have the nerve to ask the snotty spade bartender for one of the alien liquors behind the bar. It was dim. I was trying to stare everywhere at once without showing it. I remember those white boneheads – Lyrans, that is. And a mess of green veiling I decided was a multiple being from some place. I caught a couple of human glances in the bar mirror. Hostile flicks. I didn't get the message, then.

'Suddenly an alien pushed right in beside me. Before I could get over my paralysis, I heard this blurry voice:

"You air a futeball enthusiash?'

'An alien had spoken to me. An *alien*, a being from the stars. Had spoken. To me.

'Oh, god, I had no time for football, but I would have claimed a passion for paper-folding, for dumb crambo – anything to keep him talking. I asked him about his home-planet sports, I insisted on buying his drinks. I listened raptly while he spluttered out a play-by-play account of a game I wouldn't have turned a dial for. The 'Grain Bay Pashkers.' Yeah. And I was dimly aware of trouble among the humans on my other side.

'Suddenly this woman – I'd call her a girl now – this girl said something in a high nasty voice and swung her stool into the arm I was holding my drink with. We both turned around together.

'Christ, I can see her now. The first thing that hit me was *discrepancy*. She was a nothing – but terrific. Transfigured. Oozing it, radiating it.

'The next thing was I had a horrifying hard-on just looking at her.

'I scrooched over so my tunic hid it, and my spilled drink trickled down, making everything worse. She pawed vaguely at the spill, muttering.

'I just stared at her trying to figure out what had hit me. An ordinary figure, a soft avidness in the face. Eyes heavy, satiated-looking. She was totally sexualized. I remember her throat pulsed. She had one hand up touching her scarf, which had slipped off her shoulder. I saw angry bruises there. That really tore it; I understood at once those bruises had some sexual meaning.

'She was looking past my head with her face like a radar dish. Then she made an 'ahhhhh' sound that had nothing to do with me and grabbed my forearm as if it were a railing. One of the men behind her laughed. The woman said, 'Excuse me,' in a ridiculous voice and slipped out behind me. I wheeled around after her, nearly upsetting my football friend, and saw that some Sirians had come in.

'That was my first look at Sirians in the flesh, if that's the word. God knows I'd memorized every news shot, but I wasn't prepared. That tallness, that cruel thinness. That appalling alien arrogance. Ivory-blue, these were. Two males in immaculate metallic gear. Then I saw there was a female with them. An ivory-indigo exquisite with a permanent faint smile on those bone-hard lips.

'The girl who'd left me was ushering them to a table. She reminded me of a goddamn dog that wants you to follow it. Just as the crowd hid them, I saw a man join them too. A big man, expensively dressed, with something wrecked about his face.

'Then the music started and I had to apologize to my furry friend. And the Sellice dancer came out and my personal introduction to hell began.'

The red-haired man fell silent for a minute enduring self-pity. Something wrecked about the face, I thought; it fit.

He pulled his face together.

'First I'll give you the only coherent observation of my entire evening. You can see it here at Big Junction, always the same. Outside of the Procya, it's humans with aliens, right? Very seldom aliens with other aliens. Never aliens with humans. It's the humans who want in.'

I nodded, but he wasn't talking to me. His voice had a druggy fluency.

'Ah, yes, my Sellice. My first Sellice.

'They aren't really well-built, y'know, under those cloaks. No waist to speak of and short-legged. But they flow when they walk.

'This one flowed out into the spotlight, cloaked to the ground in violet silk. You could only see a fall of black hair and tassels over a narrow face like a vole. She was a mole-gray. They come in all colors. Their fur is like a flexible velvet all over; only the color changes startlingly around their eyes and lips and other places. Erogenous zones? Ah, man, with them it's not zones.

'She began to do what we'd call a dance, but it's no dance, it's their natural movement. Like smiling, say, with us. The music built up, and her arms undulated toward me, letting the cloak fall apart little by little. She was naked under it. The spotlight started to pick up her body markings moving in the slit of the cloak. Her arms floated apart and I saw more and more.

'She was fantastically marked and the markings were writhing. Not like body paint – alive. Smiling, that's a good word for it. As if her whole body was smiling sexually, beckoning, winking, urging, pouting, speaking to me. You've seen a classic Egyptian belly dance? Forget it – a sorry stiff thing compared to what any Sellice can do. This one was ripe, near term.

'Her arms went up and those blazing lemon-colored curves pulsed, waved, everted, contracted, throbbed, evolved unbelievably welcoming, inciting permutations. *Come do it to me, do it, do it here and here and here and now.* You couldn't see the rest of her, only a wicked flash of mouth. Every human male in the room was aching to ram himself into that incredible body. I mean it was *pain*. Even the other aliens were quiet, except one of the Sirians who was chewing out a waiter.

'I was a basket case before she was halfway through. . . . I won't bore you with what happened next; before it was over there were several fights and I got cut. My money ran out on the third night. She was gone next day.

'I didn't have time to find out about the Sellice cycle then, mercifully. That came after I went back to campus and discovered you had to have a degree in solid-state electronics to apply for off-planet work. I was a pre-med but I got that degree. It only took me as far as First Junction then.

'Oh, god, First Junction. I thought I was in heaven – the alien ships coming in and our freighters going out. I saw them all, all but the real exotics, the tankies. You only see a few of those a cycle, even here. And the Yyeire. You've never seen that.

'Go home, boy. Go home to your version of Burned Barn . . .

'The first Yyeir I saw, I dropped everything and started walking after it like a starving hound, just breathing. You've seen the pix of course. Like lost dreams. *Man is in love and loves what vanishes.* . . . It's the scent, you can't guess that. I followed until I ran into a slammed port. I spent half a cycle's credits sending the creature the wine they call stars' tears. . . . Later I found out it was a male. That made no difference at all.

'You can't have sex with them, y'know. No way. They breed by light or something, no one knows exactly. There's a story about a man who got hold of a Yyeir woman and tried. They had him skinned. Stories – '

He was starting to wander.

'What about that girl in the bar, did you see her again?'

He came back from somewhere.

'Oh, yes. I saw her. She'd been making it with the two Sirians, y'know. The males do it in pairs. Said to be the total sexual thing for a woman, if she can stand the damage from those beaks. I wouldn't know. She talked to me a couple of times after they finished with her. No use for men whatever. She drove off the P Street bridge. . . . The man, poor bastard, he was trying to keep that Sirian bitch happy single-handed. Money helps, for a while. I don't know where he ended.'

He glanced at his wrist watch again. I saw the pale bare place where a watch had been and told him the time.

'Is that the message you want to give Earth? Never love an alien?'

'Never love an alien – ' He shrugged. 'Yeah. No. Ah, Jesus, don't you see? Everything going out, nothing coming back. Like the poor damned Polynesians. We're gutting Earth, to begin with. Swapping raw resources for junk. Alien status symbols. Tape decks, Coca-Cola, Mickey Mouse watches.'

'Well, there is concern over the balance of trade. Is that your message?'

'The balance of trade,' he rolled it sardonically. 'Did the Polynesians have a word for it, I wonder? You don't see, do you? All right, why are you here? I mean *you*, personally. How many guys did you climb over – '

He went rigid, hearing footsteps outside. The Procya's hopeful face appeared around the corner. The red-haired man snarled at him and he backed out. I started to protest.

'Ah, the silly reamer loves it. It's the only pleasure we have left. . . . Can't you see, man? That's *us*. That's the way we look to them, to the real ones.'

'But – '

'And now we're getting the cheap C-drive, we'll be all over just like the Procya. For the pleasure of serving as freight monkeys and junction crews. Oh, they appreciate our ingenious little service stations, the beautiful star folk. They don't *need* them, y'know. Just an amusing convenience. D'you know what I do here with my two degrees? What I did at First Junction. Tube cleaning. A swab. Sometimes I get to replace a fitting.'

I muttered something; the self-pity was getting heavy.

'Bitter? Man, it's a *good* job. Sometimes I get to talk to one of them.' His face twisted. 'My wife works as a – oh, hell, you wouldn't know. I'd trade – correction, I have traded – everything Earth offered me for just that chance. To see them. To speak to them. Once in a while to touch one. Once in a great while to find one low enough, perverted enough to want to touch me. .'

His voice trailed off and suddenly came back strong.

'And so will you!' He glared at me. 'Go home! Go home and tell them to quit it. Close the ports. Burn every god-lost alien thing before it's too late! That's what the Polynesians didn't do.'

'But surely – '

'But surely be damned! Balance of trade – balance of *life*, man. I don't know if our birth rate is going, that's not the point. Our soul is leaking out. We're bleeding to death!'

He took a breath and lowered his tone.

'What I'm trying to tell you, this is a trap. We've hit the supernormal stimulus. Man is exogamous – all our history is one long drive to find and impregnate the stranger. Or get impregnated by him; it works for women too. Anything different-colored, different nose, ass, anything, man *has* to fuck it or die trying. That's a drive, y'know, it's built in. Because it works fine as long as the stranger is human. For millions of years that kept the genes circulating. But now we've met aliens we can't screw, and we're about to die trying. . . . Do you think I can touch my wife?'

'But –'

'Look. Y'know, if you give a bird a fake egg like its own but bigger and brighter-marked, it'll roll its own egg out of the nest and sit on the fake? That's what we're doing.'

'We've been talking about sex so far.' I was trying to conceal my impatience. 'Which is great, but the kind of story I'd hoped – '

'Sex? No, it's deeper.' He rubbed his head, trying to clear the drug. 'Sex is only part of it – there's more. I've seen Earth missionaries, teachers, sexless people. Teachers – they end cycling waste or pushing floaters, but they're hooked. They stay. I saw one fine-looking old woman; she was servant to a Cu'ushbar kid. A defective – his own people would have let him die. That wretch was swabbing up its vomit as if it was holy water. Man, it's deep . . . some cargo-cult of the soul. We're built to dream outwards. They laugh at us. They don't have it.'

There were sounds of movement in the next corridor. The dinner crowd was starting. I had to get rid of him and get there; maybe I could find the Procya.

A side door opened and a figure started towards us. At first I thought it was an alien and then I saw it was a woman wearing an awkward body-shell. She seemed to be limping slightly. Behind her I could glimpse the dinner-bound throng passing the open door.

The man got up as she turned into the bay. They didn't greet each other.

'The station employs only happily wedded couples,' he told me with that ugly laugh. 'We give each other . . . comfort.'

He took one of her hands. She flinched as he drew it over his arm and let him turn her passively, not looking at me. 'Forgive me if I don't introduce you. My wife appears fatigued.'

I saw that one of her shoulders was grotesquely scarred.

'Tell them,' he said, turning to go. 'Go home and tell them.' Then his head snapped back toward me and he added quietly, 'And stay away from the Syrtis desk or I'll kill you.'

They went away up the corridor.

I changed tapes hurriedly with one eye on the figures passing that open door. Suddenly among the humans I caught a glimpse of two sleek scarlet shapes. My first real aliens! I snapped the recorder shut and ran to squeeze in behind them.

Rose White

THE (INFINITE) LIBRARY OF PORN: STORAGE AND ACCESS

Decades ago, Jorge Luis Borges wrote about infinite libraries and perfect memory with the slightly sad air of someone who'd seen those things and knew their faults. Today we work toward infinite libraries and perfect memory with little heed for the possible consequences. How could it be bad to have everything possible stored? To remember everything? I don't know that it will be bad, but I do know that it will be different from our current lives of loss and forgetting. Right now, storing pornography causes problems even for people who have nothing especially perverted to hide: a collection of pornography gets to the heart of what it means to be a private individual. As we move from mass media to individually produced media, from edited collections of porn (magazines, commercially produced films) to individual snapshots and youtube clips and stored bittorrents, the particularity of a collection of porn will be testimony to its owner's private set of tastes.

Of course, it has always been a pain to store pornography - and so we have the cultural trope of a stash of magazines 'under the mattress' or in a box hidden in the closet. But as the sex industry shifts toward digital publication at every level, we might imagine that mere storage will become a problem of the past, or, at least, a problem related to legacy materials (books, magazines, videos, comic books, photographs, etc.). Cheap, massive storage media means no more problem, right?

The more things change, the more they stay the same – every new technology disrupts its society, creating unexpected problems as well as the hoped-for benefits.

Our information overload, which feels like such a contemporary problem, is not new to the millennial generation or even the baby boomers; it wasn't even new to Vannevar Bush when he wrote 'As We May Think' in 1945.

When Frances Yates wrote the *Art of Memory* in 1966 she described the 'memory castles' used by medievals and the ancients to keep track of information that was too unwieldy to be remembered by simpler means. So the organization and storage and retrieval of the information that humans produce has been a problem for millennia. We are inherently creative! We make words and images and we like to share them! (CEILING CAT IS WATCHING YOU MASTURBATE!)

So what's so special about sexuality as a topic for 'serious study?'

Ceiling Cat, http://www.ceilingcat.com/img/ceilingcat.jpg

Academically, it really is a special case. If you talk about pornography, and you use your real name, you are marked. My colleague Audacia Ray has written about her problems being taken seriously in academia. My own mentor told me that I should keep the 'porn and sex-related writing' off my CV, and use a pseudonym for it – but my friend Robert Lawrence said that maybe it's a little too late for me, and so perhaps I should use a pseudonym for my sociological writings.

I wanted to mention those anecdotes specifically because they point to the general problem: embarrassment about subject matter complicates an already incredibly difficult information management issue: if you cannot talk about what it is you are storing in any more detail than with a collective noun (pr0n!) then none of the existing solutions can even be brought to bear on the problem!

The canonical legacy porn storage problem was the stash of porn under the mattress or on a closet shelf or somewhere in the garage. There are some assumptions implicit in this ideal porn collection: it is kept by an adult man; it's relatively small; it may be replenished with new material, but then old material is thrown away to keep the size constant; in the United States, despite being almost certainly legal and limited to newsstand purchases, this collection would still be private, and kept physically hidden from a girlfriend or wife or children.

Although of course pornographic magazines are still sold, this type of porn collection is increasingly rare - digital porn has completely overwhelmed the market. With this shift, there is no longer a single 'ideal type' of collection – there are as many different collections of pornography as there are individuals who save files of it.

The privacy implications of this are profound, so I want to pause here. The choice to store pornography causes problems even for people with nothing to hide beyond their activity as sexual beings. The act of curating a collection of pornography, the *particularity* of a set of pornographic images, testifies to an individual's predilections. For instance, if your partner sees some porn magazines you own and gets upset because there are lots of big-breasted women, you might object that you have no control over what goes into the magazines, and you just wanted a lad-mag, and that's what one gets in them. But if your partner finds your porn files and sees that you've collected two gigs of images of different naked people who all look like your ex, you're going to have a lot of explaining to do – clearly you were able to exercise some volition in curating that collection of images.

Violet Blue spoke at *Arse Elektronika* in 2007 on this topic of privacy and our 'erotic fingerprints'. How we organize any of our data fingerprints us, but perhaps nowhere more uniquely than if we catalogue or record our sexual desires. She pointed out, and I agree, that sometimes we are only honest with our own selves about those private desires – there are things that go unspoken even to beloved partners – and that this is completely okay, and part of being human. We can request our right to privacy in these matters as well as all others – and without embarrassment. As we should all know: just because someone has a desire for privacy does not imply that they are doing anything illegal, or illicit, or even merely morally wrong. It's okay for some things to be private!

HOWEVER!

Embarrassment, and privacy concerns, and often overtrumped legal worries all intermingle to make pornography a weird train wreck of a topic to study. Few people want to say what they mean with any specificity, and people squirm (and not in a good way!) if you try to make them be more specific.

Librarians, for instance, usually love to tag items and provide exhaustive metadata for materials in their collections – providing access to media is their *job*. However, the Library of Congress cataloging system is an Epic Fail on porn. They are not strong on subculture material in general, but there seems to be a tacit agreement about pornographic holdings: depending on the library, if you know that what you want is in the system, then you can request it and withdraw it from the library. If you don't know the exact title, you will have a *very* hard time doing subject searches. There isn't enough room for specificity in the cataloging system itself. Furthermore, books and videos 'of a pornographic nature' are sometimes held in closed stacks, and not available for interlibrary loan. Protecting library holdings from damage or loss through obscurity seems like bad practice to me!

The desire to obscure what is stored makes the language of talking about pornography, even *completely legal* pornography, sometimes seem closer to the diction of corporate espionage and international spying.

Mostly when we store information we want to be able to retrieve it easily, and we'd be happy to have a straightforward cataloging methodology. The latter is why a music freak I know has his many thousands of CDs completely alphabetized, and has a few genres broken out and alphabetized separately.

My preliminary questioning of people who've been open to talking to me about their porn collections has led to some startling anecdotes, however. That same music freak stores his pornographic images on about a dozen CDs, which are simply labeled 'porn 2000', 'porn early 2001' and so on. On the CDs, the folders sometimes mention a website that the images come from, and sometimes a specific model, and sometimes just a month that they were downloaded. He joked to me that sometimes instead of looking for new porn he'd go back and say, 'Ah, fall 2003, sure, that was a good season for porn!' The CD storage he said was to get stuff off a too-small hard drive, but not to hide it from his wife or keep it off his main computer. When I asked why he didn't catalogue it as thoroughly as his music, he said, 'Oh, well, I'd feel like a pervert if I did that, right? That would be paying too much attention to it.' So, to be clear: He was embarrassed to spend any effort organizing the porn, because of what it might mean about his identity – even though no one else had ever had any knowledge of or interest in this collection.

Another person I spoke with did have a girlfriend whom he wanted to hide his porn from. He described it as being straightforward 'pictures and videos of pretty naked girls, nothing too kinky' and gave *abbywinters.com* as an example of the sort of site he likes. He said he sometimes subscribes for a month, downloads lots, and then unsubscribes, which he likes doing because then he doesn't have the site flashing at him and trying to sell him things when he just wants to look at porn. He has a few sites he likes and trusts to not rip off his banking information, so he can switch around when he wants to look for new stuff. When I asked, though, he mentioned that his interest in novelty wasn't related to having 'run out of porn,' and that in fact he was sure he had more than a 'lifetime's worth' of porn stored already. He said he didn't see any reason to *not* have lots of porn, since it's trivial to save the files, and storage is cheap. However, he worries that his girlfriend will find out that he has this much porn and think he is a 'freak'. So he has hidden the files on his desktop computer. He uses Windows, and he put a hidden folder in the '\Windows' directory with a dummy name, then in *that* folder there are folders named with letters and numbers that are all fake, and in one of *those* is his enormous directory of porn. 'My girlfriend would never look in there because she would be afraid of breaking something,' he said.

Leaving aside one fellow who told me he stores his porn 'in the cloud' (by which he meant there's so much porn online that he doesn't need to save any, he can just have a look around the web whenever he'd like to see some) all the other people I have spoken with personally also save pornographic text files, images, and videos to their own computers. My sample is both small and unusual, because they are people who were comfortable talking to a friend about their use and storage of pornography, so I cannot yet make any really generalizable statements about them. I do find it very interesting, though, that so far the people I have spoken with have all had a strong interest in curating these collections because the context of online pornography detracted so much from their enjoyment.

One woman complained that it is nearly impossible to find images of the sex acts that she wants to see that are not part of what is, for her, an unpleasant or even violent narrative. She likes fisting, but in the narrative context of pornography, this is equivalent to scat and heavy bondage and sadomasochism. She sees her forays into the world of online pornography as 'hunting and gathering missions' in a hostile environment.

The people I've spoken with are using flat file structures to organize their pornography. No one is using a relational database, even though in at least two cases the amount of data and the cross-referencing involved seems like it would make that appropriate. I'll be very interested to see, once I start talking to more people, to find individuals who have taken on more thorough cataloging projects, and to hear how they've handled those.

I have had a tiny hint of one thing that's going on online: lots of people are using *Flickr*. It's a brilliant hack of *Flickr's* setup: you can get a free account, and without posting any pictures of your own, you can favorite other people's photos. Some people make separate accounts for each special interest they have – like feet, or socks, or cleavage, or brunettes, and so their favorites pages are full of just those images, hundreds of them. If they like, they can also join groups on *Flickr* that cater to the same interest, and find even more photos. Instead of keeping the photos offline (or in addition), the user can access them from any browser, all sorted by interest, and tagged by the community. It's a ginormous overlapping fractal fetish library!

I wonder whether *Flickr's* API itself, or even just its general model, will offer a way for lifelogging to take off. Futurists like Clay Shirky and Charlie Stross and Kevin Kelly have been talking about the dangers and wonders of lifelogging for quite awhile now, but it's not quite present reality yet. But all it takes is switching on the cellphone cameras or the *EEE PC* built-ins 24/7, and suddenly I'm looking at a congress hall full of cam-whores. Then you'll have all this wonderful footage, some of which I'm sure will be fantastically sexy and worth saving, and BOOM: you've got yet another datastream to save and manage.

I have a project brewing for next year's *Arse Elektronika* that foreshadows the day when we'll be able to indulge in fully immersive AI-driven pornographic experiences (such as texting back-and-forth with artificially intelligent SMS-bots, sending texts and photos and audio to a perfectly responding far-away 'partner'), and we'll also want some way to keep those experiences.

Without some sort of index, we're lost in an infinite library. But I don't know if we're any more lost than we once were. We used to just have less porn; we used to just document less of our lives. At least we'll have different memories to comfort us in *our* old age, that's for certain.

BIO NOTES

monochrom

monochrom is a worldwide operating collective dealing with technology, art, context hacking, and philosophy which was founded in 1993. They specialize in an unpeculiar mixture of proto-aesthetic fringe work, pop attitude, subcultural science, and political activism. Their mission is conducted everywhere, but first and foremost 'in culture-archaeological digs into the seats (and pockets) of ideology and entertainment.'

Among their projects, monochrom has started *Arse Elektronika*, has released a leftist retro-gaming project, established a one baud semaphore line through the streets of San Francisco, started an illegal space race through Los Angeles, buried people alive in Vancouver, and cracked the hierarchies of the art system with the Thomann Project. They ate blood sausages made from their own blood in order to criticize the grotesque neoliberal formation of the world economy. Sometimes they compose melancholic pop songs about dying media and host the first annual festival concerned with cocktail robotics. They also do international soul trade, political sock puppet shows, aesthetic pregnancy counselling, food catering, and - sorry to mention - modern dance.

monochrom members:

Johannes Grenzfurthner, Franz Ablinger, Harald List, Evelyn Fürlinger, Frank Apunkt Schneider, Daniel Fabry, Günther Friesinger, Anika Kronberger, Roland Gratzer.

www.monochrom.at/english/

Editors:

Johannes Grenzfurthner is an artist, writer, curator, and director. He is the founder of monochrom. He teaches art theory and aesthetical practice at the University of Applied Sciences in Graz, Austria. Tag cloud: contemporary art, activism, performance, humor, philosophy, postmodernism, media theory, cultural studies, science fiction, and the debate about copyright.

Günther Friesinger lives in Vienna and Graz as a philosopher, artist, curator, and journalist. He is member of monochrom and he works for the University of Vienna and Team Teichenberg.

Daniel Fabry is researcher, lecturer, and artist in the lofty fields of media and interaction design. He is member of monochrom and works at the University of Applied Sciences in Graz, Austria.

Thomas Ballhausen studied Comparative Literature and German at the University of Vienna. He is a lecturer at the University of Vienna and at the University of Applied Arts, head of the Studies-Department of the Austrian Film Archive, and has authored publications on film history, media theory, and popular culture.

Violet Blue is pro blogger, podcaster, vlogger and femmebot at Metblogs SF, Geek Entertainment TV, and Gawker Media's Fleshbot. And sex columnist for the San Francisco Chronicle, and a Forbes Web Celeb. She is a best-selling, award-winning author and editor of almost two dozen books; some translated into five languages.

Jason Brown is an ambient noisemaker, constellation manipulator, and paranoid historiographer. He is consigliere of Machine Project, a Los Angeles based non-profit which encourages heroic experiments of the gracefully over-ambitious. He is director of Superbunker, a framework for conducting and disseminating critical and creative research. He was a founding member of c-level, a collaborative group which focused on media, protest and play. He is acting janitor of Beta-level, an underground venue beneath Chinatown. He is an instructional technologist at Pomona College.

Reesa Brown's short story "Memory Box" will be published in fall of 2008 in Unspeakable Horror: From the Shadows of the Closet from Dark Scribe Press. Her flash fiction, "The Reap Assessors" will appear in Triangulation: Taking Flight from PARSEC Ink. Together with best-selling fantasy author Steven Brust, she blogs at Words Words Words (http://dream-cafe.com/words/).

Benjamin Cowden began working with metal during an undergraduate anthropology project in Cameroon in 1997, where he studied how Baka Pygmies turned worn machetes into utility knives. He later worked as a blacksmith and focused on utilitarian objects and furniture. When he began making sculpture in 2003, Cowden sought to maintain the physical relationship and interactivity of his earlier work. During his graduate studies at Southern Illinois University Carbondale, Benjamin created interactive mechanical devices which explored human experience and the senses. Mr. Cowden now lives in Oakland California where he continues to explore these themes using mechanics, electronics, and other media.

Cory Doctorow (craphound.com) is a science fiction author, activist, journalist and blogger - the co-editor of Boing Boing (boingboing.net) and the author of the bestselling Tor Teens novel LITTLE BROTHER. He is the former European director of the Electronic Frontier Foundation and co-founded the UK Open Rights Group. Born in Toronto, Canada, he now lives in London.

Karin Harrasser studied German Literature and History at the University of Vienna, 2001-2002 Juniorfellow at the IFK and Research Scholar at Duke University, 2005 Phd at the University of Vienna, thesis on the narratives of digital cultures of the 1980ies, 2005-2007 post-doc position at the Graduate Seminar "Codes of Violence in Changing Media" at the Humboldt-Universität Berlin. Since then employed at the Univ. of Vienna and the Humboldt-University as researcher. Recent research: a history of prosthetic knowledge (habil-project), research project on the production of gender and knowledge in museums.

Richard Kadrey is a novelist, freelance writer, and photographer based in San Francisco. Kadrey's novels are Metrophage, Kamikaze L'Amour, and Butcher Bird: A Novel Of The Dominion." Other works include collaborative graphic novels and over 50 published short stories. Kadrey's short story Carbon Copy: Meet the First Human Clone was filmed as After Amy. His non-fiction books as a writer and/or editor include The Catalog of Tomorrow (Que/TechTV Publishing, 2002), From Myst to Riven (Hyperion, 1997), The Covert Culture Sourcebook and its sequel (St. Martin's Press, New York, 1993 and 1994); Kadrey also hosted a live interview show on Hotwired in the 1990s called Covert Culture. He was an editor at print magazines Shift and Future Sex, and at online magazines Signum and Stim. He has published articles about art, culture and technology in publications including Wired, Omni, Mondo 2000, the San Francisco Chronicle, SF Weekly, Ear, Artforum, ArtByte, Bookforum, World Art, Whole Earth Review, Reflex, Science Fiction Eye, and Interzone.

Isaac Leung, who was born in Hong Kong, is an artist, curator and researcher, based in Hong Kong and the United States, whose work focuses on art, technology and sexuality. He is currently a Mphil student at the Cultural Studies Department of Lingnan University and is on the board of directors of the Hong Kong Sex Education Association and the Videotage Media Arts Organization. He received a Bachelor of Fine Arts degree at the New Media Art Department of the School of the Art Institute of Chicago and Central Saint Martins College of Art and Design and a Honorary Fellowship from the Art Institute of Chicago. He has given lectures about this project at the University of Hong Kong (Hong Kong), City University of Hong Kong (Hong Kong) and the School of the Art Institute of Chicago (USA).

Tina Lorenz is a long-time member of the Chaos Computer Club, focussing on the impact of technology on society. She also studies theatre-, film- and media research in Vienna and Munich. She is writing her master's thesis about pornographic film.

Susan Mernit (http://www.susanmernit.com) is the co-founder of People's Software Company, a start-up building interactive local community platforms, a BlogHer contributing editor on Sex & Relationships (http://www.blogher.com/topic/sex-relationships), a former exec at Yahoo, AOL, Netscape and Advance Internet, and a sex positive and feminist-identified person, writer, and troublemaker. She is also an evangelist for the 2008-09 Knight News Challenge (http://www.newschallenge.org/); talk to her about applying for grants to build open source community projects that support news, discourse and the commons.

Mela Mikes is currently living and working in Vienna. Software Test Engineer and university drop out after studying Philosophy for a while. Hobby DJ and creator of the melafesto podcasts.

Annalee Newitz is a writer who covers the collisions between technology and media, culture and science. She writes for many periodicals from Popular Science to Wired, and since 1999 has had a syndicated weekly column called Techsploitation. From 2004-2005 she was a policy analyst for the Electronic Frontier Foundation. She is the editor of io9, a blog about science fiction owned by the Gawker Media Network.

Chris Noessel is a consultant at Cooper. His industry experience ranges from owning a small, museum-focused company in Houston to working with Microsoft's futures prototyping group. He was Director of Information Architecture at USWeb/CKS (through the marchFirst and SBI acquisitions), developing their user interface practice from scratch, developing training curricula for designers and developers, and consulting to clients. He's been an entrepreneur (founding his own firm in 1995) as well as a researcher. He was a founding student at Interaction Design Institute Ivrea in 2001 and graduated with a Masters in Interaction Design in 2003.

Kit O'Connell is a author of science fiction and erotica, as well as works which combine the two. His poem, "a 24th-century reflection on emptiness" was recently published by Aberrant Dreams. Together with best-selling fantasy author Steven Brust, he blogs at Words Words Words (http://dreamcafe.com/words/).

Jens Ohlig is a long-time activist at the Chaos Computer Club in Germany. His major fields of interest are the intersections of linguistics, literature, and computer sciences. According to Wikipedia criteria he is irrelevant, but thinks he is in good company. The presentation is the result of an ongoing research together with Svenja Schröder, Ph.D. student in computer science at the University of Duisburg-Essen, Germany.

Stephane Perrin aka 23N! (zeni) started to build his own musical devices in 2005. In early 2007 he started the band Droise with Kakawaka, another noise musician, and Yosei, a drummer. Together they played numerous shows in Japan and did a tour in Europe in Fall 2007. 23N! is now continuing as No Fork Droise with drummer Yosei and has recently started to play solo shows that are a mix of visual performance, cosplay and noise, always with homemade instruments, noise toys and effects. In addition to his musical activities, 23N! has started various media art projects. He is running the noise music label Even Stilte and regularly organizes live shows in Tokyo.

Bonni Rambatan is an independent cultural researcher, theorist, and blogger. His primary field of research is the role of technology in shaping contemporary human subjectivity, sexuality, and society. Rambatan aims to develop a critical approach to society and politics with his unique blend of Lacanian psychoanalysis and informatics and new media studies, interrogating our experience with the computer monitor and how it alters our notion of subjectivity and our relations to local and global politics and the market. Rambatan's blog can be found at http://posthumanmarxist.wordpress.com.

Thomas S. Roche is a writer of fiction - particularly erotic fiction, crime and speculative fiction - and nonfiction - particularly interviews and commentary on erotic expression and other aspects of human sexuality. His more than 400 professionally published short stories have appeared in a wide variety of magazines, anthologies and websites; 2008 will mark his fourth inclusion in Susie Bright's Best American Erotica series. He has been a staff educator for nearly a decade with the nonprofit educational service San Francisco Sex Information (www.sfsi.org), and is currently the Managing Editor of Eros Zine (www.eros-zine.com). His previous work has included staff writing positions at www.gettingit.com, www.gothic.net. com, www.libida.com, www.13thstreet.com, and www.goodvibes.com; he served as Marketing Manager of Good Vibrations from 2001-2004. He is the author or editor of eleven published books and coordinates the San Francisco edition of the international burlesque figure drawing salon, Dr. Sketchy's Anti-Art School (www.drsketchy.com). His homepage is www.skidroche.com.

Bonnie Ruberg is a sex and games journalist. She writes for a number of publications, including The Village Voice and Forbes.com.

Rudy Rucker is computer scientist and science fiction author, and is one of the founders of the cyberpunk literary movement. The author of both fiction and non-fiction, he is best known for the novels in the Ware Tetralogy, the first two of which (Software and Wetware) both won Philip K. Dick Awards. At present he edits the science fiction webzine Flurb.

Mae Saslaw is a writer who lives and works in Brooklyn, New York. Her body of work includes short contemporary fiction as well as literary and cultural criticism. Last year she began an extensive hypertext called The Machine-Space Project, which examines the Internet as space of opposition between revolutionary and capitalist-hegemonic forces. Her other academic interests include Baroque history and pop politics. She spends time assisting and collaborating with writers and artists in New York, and she studies at the Pratt Institute.

Sharing is Sexy (DJ Lotus, Kelly Lovemonster, J Bird, Scruffy Eudora): In looking at most mainstream porn, we don't see ourselves and our friends. In response, we are creating the porn that we want to see. In our project, we are sharing our bodies with you willingly. We're in control, do not change the channel. Our collective practices consensus. Everyone has an equal voice in what we do and how we do it. We are examining the question of open source porn while respecting the work of sex workers. We're in this to create change. Beyond the website, we are also organizing events that advocate sex-positive ideas, having porn watching nights, hosting burlesque shows and making zines.

SiS is a collaborative open source porn laboratory. We are a group of queer people, transgender people and people with othered bodies coming together to create a site for free porn that is licensed under Creative Commons BY-NC-SA. We are creating our own porn using photos, video, and writing. SiS is polyamorous so, we are open to new members and looking to collaborate.

Nathan Shedroff is one of the pioneers in Experience Design, an approach to design that encompasses multiple senses and requirements as well as related fields, Interaction Design and Information Design. His speaking, books, teaching, and projects all support this new direction of design. Part designer, part entrepreneur, his skills lend themselves to strategic thinking and design for companies who want to exploit the strengths of experience media in order to build better experiences for their customers and themselves in a variety of media, including: print, digital, online, and product design. Growing-up in Silicon Valley has given him an entrepreneurial outlook. He currently lives in San Francisco where the climate, culture, and industry make it easy to have an esoteric and amorphous title like Experience Strategist and actually make a living.

James Tiptree, Jr. (August 24, 1915 - May 19, 1987) was the pen name of American science fiction author Alice Bradley Sheldon, used from 1967 to her death. She also occasionally wrote under the pseudonym Raccoona Sheldon (1974-77). Tiptree/Sheldon was most notable for breaking down the barriers between writing perceived as inherently "male" or 'female' it was not publicly known until 1977 that James Tiptree, Jr. was a woman.

Allen Stein is co-founder of www.sexmachinecams.com and a teledildonics pioneer. Inventor of the first-to-market teledildonics sex machine-thethrillhammer-he has been involved in the science for 8 years. The initial popularity of thethrillhammer propelled the company into the adult industry designing various luxury pleasure devices and producing content for the adult industry for consumer consumption. The company offers retail and web based entertainment products for income thus providing a means to continue his research into female sexual response. He has designed and prototyped various immersive interfaces and VR simulations to advance the way humans interact sexually with machines.

Viviane (http://www.thesexcarnival.com) is an NYC based librarian, sex nerd, and blog mentor. She started the sex positive group blog Viviane's Sex Carnival in 2005. She was the editor of TGP.com. As a Fleshbot contributor she contributor a weekly sex blog roundup. In 2006 she organized the regular gathering of sex bloggers in NY aka the Perverts Saloon. She has presented at Dark Odyssey and Sex 2.0 about blog setup and promotion.

Jörg Vogeltanz has a stage design diploma (1988-92, University of Arts, Graz). He is freelance artist in different fields of interests and techniques (painting, graphic design and layout, illustration, graphic novels/comics, video clip and commercials directing, film/video artwork, animation etc.). He is lecturer for freehand drawing (FH Joanneum, Graz) and freelance online journalist.

Rose White is a sociology PhD student at Graduate CenterCUNY, in Manhattan, but lives in Brooklyn because that's where all the action is. Her doctoral work is on how bright people break rules (in other words, deviance and technology!)-tell her what you want studied and maybe she'll get on that. In 1997 she had a story anthologized in Best American Erotica, which means it's about time for her to write some new porn.

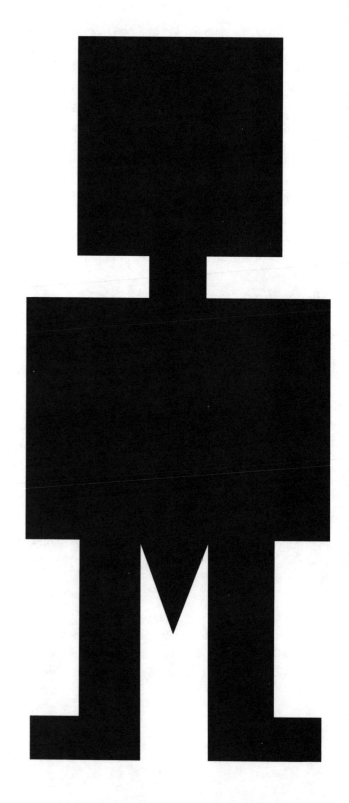

253

RE/Search Catalog

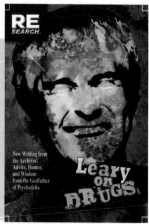

NEW! RE/SEARCH CLASSICS IN DELUXE HARDBACK

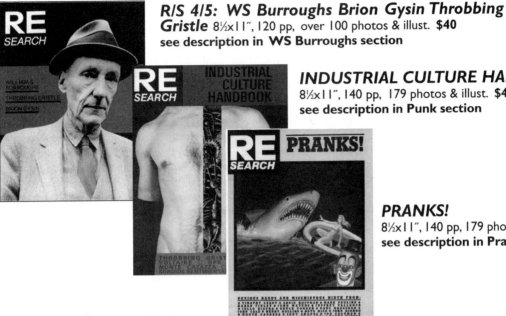

R/S 4/5: WS Burroughs Brion Gysin Throbbing Gristle 8½x11", 120 pp, over 100 photos & illust. **$40**
see description in **WS Burroughs section**

INDUSTRIAL CULTURE HANDBOOK 8½x11", 140 pp, 179 photos & illust. **$40**
see description in **Punk section**

PRANKS! 8½x11", 140 pp, 179 photos & illust. **$40**
see description in **Pranks section**

SPECIAL OFFER
ALL THREE CLASSIC HARDBACKS FOR JUST $90 PLUS SHIPPING

MEDIA

LOUDER FASTER SHORTER punk film by Mindaugis Bagdon now on DVD.

San Francisco, March 21, 1978. In the intense, original punk rock scene at the Mabuhay Gardens (the only club in town which would allow it), the AVENGERS, DILS, MUTANTS, SLEEPERS and UXA played a benefit for striking Kentucky coal miners ("Punks Against Oppression!"). One of the only surviving 16mm color documents of this short-lived era, *LOUDER FASTER SHORTER* captured the spirit and excitement of "punk rock" before revolt became style. Filmmaker Mindaugis Bagdon was a member of *Search & Destroy*, the publication which chronicled and catalyzed the punk rock "youth culture" rebellion of the late '70s. "Exceptionally fine color photography, graphic design and editing."—S.F. International Film Festival review, 1980. 20 minutes. DVD limited edition made December 2007 of only 500 copies: **$20.**

COUNTER CULTURE HOUR INTERVIEWS AVAILABLE ON DVD:
LYLE TUTTLE, GENESIS P-ORRIDGE, ROBÖXOTICA 2006
AND MORE! PLEASE EMAIL FOR DETAILS.

Visit www.researchpubs.com to see more of our books, CDs and videos! Jean-Jacques Perrey! Ken Nordine! PRANKS! Swing! & more!

PUNK & D.I.Y.

PUNK '77: an inside look at the San Francisco rock n' roll scene, 1977 by James Stark

Covers the beginnings of the S.F. Punk Rock scene through the Sex Pistols' concert at Winterland in Jan., 1978, in interviews and photographs by James Stark. James was among the many artists involved in early punk. His photos were published in *New York Rocker*, *Search & Destroy* and *Slash*, among others. His posters for Crime are classics and highly prized collectors' items. Over 100 photos, including many behind-the-scenes looks at the bands who made things happen: Nuns, Avengers, Crime, Screamers, Negative Trend, Dils, Germs, UXA, etc. Interviews with the bands and people early on the scene give intimate, often darkly humorous glimpses of events in a *Please Kill Me* (Legs McNeil) style.
"The photos themselves, a generous 115 of them, are richly satisfying. They're the kind of photos one wants to see..."—Puncture. "I would recommend this book not only for old-timers looking for nostalgia, but especially to young Punks who have no idea how this all got off the ground, who take today's Punk for granted, to see how precarious it was at birth, what a fluke it was, and to perhaps be able to get a fresh perspective on today's scene needs..."—MAXIMUMROCKNROLL 7½x10¼", 98 pp, 100+ photos, on archival art paper. PB, **$20**

SEARCH & DESTROY: The Complete Reprint (2 big 10x15" volumes)
"The best punk publication ever"—Jello Biafra

Vol. 1
out of print

Facsimile editions (at 90% size) include all the interviews, articles, ads, illustrations and photos. Captures the enduring revolutionary spirit of punk rock, 1977-1978. Vol. I contains an abrasive intro-interview with Jello Biafra on the history and future of punk rock. Published by V. Vale before his RE/Search series, *Search & Destroy* is a definitive, first-hand documentation of the punk rock cultural revolution, printed as it happened! Patti Smith, Iggy Pop, Ramones, Sex Pistols, Clash, DEVO, Avengers, Mutants, Dead Kennedys, William S. Burroughs, J.G. Ballard, John Waters, Russ Meyer, and David Lynch (to name a few) offer permanent inspiration and guidance. First appearance of Bruce Conner Punk photos. 10x15", 148pp, Only 200 copies left of Volume Two: **$19.95 (Volume One is long out-of-print)**

Some of the Search & Destroy original tabloid issues are still available. Please call or email for information.

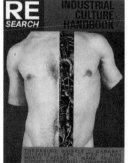

INDUSTRIAL CULTURE HANDBOOK DELUXE HARDBACK
This book is a secret weapon—it provided an educational upbringing for many of the most radical artists practicing today! The rich ideas of the *Industrial Culture* movement's performance artists and musicians are nakedly exposed: *Survival Research Laboratories, Throbbing Gristle, Cabaret Voltaire, SPK, Non, Monte Cazazza, Johanna Went, Sordide Sentimental, R&N, & Z'ev*. **Topics include:** brain research, forbidden medical texts & films, creative crime & *interesting* criminals, modern warfare & weaponry, neglected gore films & their directors, psychotic lyrics in past pop songs, and *art brut*. Limited Edition of only 1000 hardbacks on glossy paper.
8½x11", 140 pp, 179 photos & illust. PB, **$40**

ZINES Vol. 1 & 2
The Punk Rock Principle of "DO-IT-YOURSELF" (D-I-Y) inspired the creation of "ZINES": handmade self-publications by creative individuals on topics ideally against status quo thinking and ideas. Zines such as Murder Can Be Fun, Beer Frame, Crap Hound, Thrift Score, Bunny Hop, Fat Girl, Housewife Turned Assassin gleefully show the satisfactions to be had by "publishing it yourself." These two books will inspire and provoke readers to become publishers, and have been used in college classes as textbooks. Both books heavily illustrated with photographs and illustrations from Zinemakers' lives & works. **Zines 1:** 184 pp, PB, **$18.99. Zines 2:** 148 pp, PB, **$14.99. *GET BOTH* for $20.00 (plus shipping for two items).**

www.researchpubs.com • (415) 362-1465 • info@researchpubs.com

BODY MODIFICATION AND S&M

RE/Search 12: MODERN PRIMITIVES The *New York Times* called this "the Bible of the underground tattooing and body piercing movement." *Modern Primitives* launched an entire '90s subculture. Crammed with illustrations & information, it's now considered a classic. The best texts on ancient human decoration practices such as tattooing, piercing, scarification and more. 279 eye-opening photos and graphics; 22 in-depth interviews with some of the most colorful people on the planet. "Dispassionate ethnography that lets people put their behavior in its own context."—*Voice Literary Supplement* "The photographs and illustrations are both explicit and astounding . . . provides fascinating food for thought. —*Iron Horse* 8½x11", 212 pp, 279 photos and illus, PB. Great gift! **$19.50**

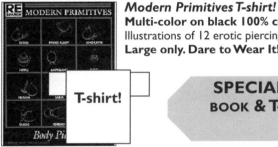

Modern Primitives T-shirt!
Multi-color on black 100% cotton T-shirt
Illustrations of 12 erotic piercings and implants. **Xtra Large only. Dare to Wear It! $16.**

SPECIAL OFFER: MODERN PRIMITIVES
BOOK & T-SHIRT GIFT-PACK—ONLY $29.00
PLUS SHIPPING.

Confessions of Wanda von Sacher-Masoch Married for 10 years to Leopold von Sacher-Masoch (author: *Venus in Furs* & many other novels) whose whip-and-fur bedroom games spawned the term "masochism," Wanda's story is a feminist classic from 100 years ago. She was forced to play "sadistic" roles in Leopold's fantasies to ensure the survival of herself & their 3 children–games which called into question who was the Master and who the Slave. Besides being a compelling story of a woman's search for her own identity, strength and, ultimately, complete independence, this is a true-life adventure story–an odyssey through many lands peopled by amazing characters. Here is a woman's consistent unblinking investigation of the limits of morality and the deepest meanings of love. "Extravagantly designed in an illustrated, oversized edition that is a pleasure to hold. It is also exquisitely written, engaging and literary and turns our preconceptions upside down."—*L.A. Reader* 8½x11", 136 pp, illustrated, PB. **$20**

The Torture Garden by Octave Mirbeau This book was once described as the "most sickening work of art of the nineteenth century!" Long out of print, Octave Mirbeau's macabre classic (1899) features a corrupt Frenchman and an insatiably cruel Englishwoman who meet and then frequent a fantastic 19th century Chinese garden where torture is practiced as an art form. The fascinating, horrific narrative slithers deep into the human spirit, uncovering murderous proclivities and demented desires. "Hot with the fever of ecstatic, prohibited joys, as cruel as a thumbscrew and as luxuriant as an Oriental tapestry. Exotic, perverse . . . hailed by the critics."—Charles Hanson *Towne* 8½x11", 120 pp, 21 mesmerizing photos. **PB: $25. Rare Hardcover (edition of only 100; treat yourself!): $40**

Bob Flanagan Super Masochist Born 1952 and deceased in 1996, Bob grew up with Cystic Fibrosis, and discovered extended S&M practices as a secret, hand-picked pathway towards life extension. In flabbergastingly detailed interviews, Bob described his sexual practices and his relationship with long-term partner and Mistress, the artist Sheree Rose. Through his insider's perspective we learn about branding, piercing, whipping, bondage and ingenious, improvised endurance trials. Includes photographs by Sheree Rose. This book "inspired" a movie, which used many of the questions found in this book. "...an elegant tour through the psychic terrain of SM." -- Details Magazine 8½x11", 136 pp, illustrated, PB. **$20.00**

PRANKS

RE/Search 11: PRANKS! A prank is a "trick, a mischievous act, a ludicrous act." Although not regarded as poetic or artistic acts, pranks constitute an art form and genre in themselves. Here pranksters challenge the sovereign authority of words, images and behavioral convention. This iconoclastic compendium will dazzle and delight all lovers of humor, satire and iron. "I love this book. I thought I was the only weirdo out there, but this book inspires me to be weirder. I pick it up weekly, even though I've read it from covr to cover many times. Still cracks me up." (reader) "The definitive treatment of the subject, offering extensive interviews with 36 contemporary tricksters including Henry Rollins, Abbie Hoffman, Jello Biafra, SRL, Karen Finley, John Waters, DEVO … from the Underground's answer to Studs Terkel."—*Washington Post* 8½x11″, 240 pp, 164 photos & illustrations, PB, **$25, Deluxe HB $40.**

PRANKS 2 *"Pranks woke me up from a deep slumber. It's as if a demolition crew had a party in my brain." (reader)* An extended underground of surrealist artists like The Suicide Club, Billboard Liberation Front, DEVO, John Waters, Lydia Lunch & Monte Cazazza give inspiring tales of mirth and conceptual mayhem. Includes Internet pranks, art pranks, prank groups and more! 8'x10″, 200 pp,, many photos & illustrations, PB, glossy paper, 8x10″, **$20.**

Mr Death T-Shirt by ManWoman. Black & yellow on White 100% heavyweight cotton T-Shirt. Sizes: S,M,L,XL. Limited edition, 50 copies: **$25.**

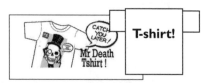

TWO BY DANIEL P. MANNIX

MEMOIRS OF A SWORD SWALLOWER Not for the faint-of-heart, this book will GROSS SOME PEOPLE OUT and delight others. "I probably never would have become America's leading fire-eater if Flamo the Great hadn't happened to explode that night …" So begins this true story of life with a traveling carnival, peopled by amazing characters—the Human Ostrich, the Human Salamander, Jolly Daisy, etc.—who commit outrageous feats of wizardry. One of the only *authentic* narratives revealing the "tricks" (or rather, painful skills) involved in a sideshow, and is invaluable to those aspiring to this profession. OVER 50 RARE PHOTOS taken by Mannix in the 1930s. 8½x11″, 128 pp, 50+ photos, index, PB, **$20**
Signed copies available for only $30

FREAKS: We Who Are Not As Others
Amazing Photos! This book engages the reader in a struggle of wits: Who is the freak? What is normal? What are the limits of the human body? A fascinating, classic book, based on Mannix's personal acquaintance with sideshow stars such as the Alligator Man and the Monkey Woman. Read all about the notorious love affairs of midgets; the amazing story of the Elephant Boy; the unusual amours of Jolly Daisy, the fat woman; hermaphrodite love; the bulb-eating Human Ostrich, etc. **Put this on your coffee table and watch the fun!** 8½x11″, 124 pp, 88 photos. PB. **$20** Author died in 1997. **Signed, hardbound copies available for $50**

J.G. BALLARD

J.G. Ballard: Quotes Amazing, provocative quotes from J..G. Ballard illuminating the human condition, arranged by topic. Edited by V. Vale with Mike Ryan. Dozens of gorgeous photos by Ana Barrado, Charles Gatewood and others. "Ballard understands the transformation technology can effect on human desire." (Observer) ISBN 1-889307-12-2, 416 pages, index, 5" x 7", gloss paper, **$20.** Limited AUTOGRAPHED Flexibind Edition of only 250 copies, signed by J.G. Ballard himself, only **$75.** Also, unsigned Library Flexibind Editions (only 100 printed) available, only **$35.**

J.G. Ballard Conversations

British luminary, J.G. Ballard converses with V. Vale, Mark Pauline, Graeme Revell, David Pringle and other forward thinkers. Photographs by Ana Barrado, Charles Gatewood and others. Some topics: Sex, technology, the future, plastic surgery, child-raising, *Empire of the Sun*. Index, book recommendations, and interviews are also included.
ISBN 1-889307-13-0, 360 pages, index, 5" x 7", gloss paper, **$20**

RE/Search 8/9: J.G. Ballard J.G. Ballard is our finest living visionary writer. His classic, *CRASH* (made into a movie by David Cronenberg) was the first book to investigate the psychopathological implications of the car crash, uncovering our darkest sexual crevices. He accurately predicted our media-saturated, information-overloaded environment where our most intimate fantasies and dreams involve pop stars and other public figures. Intvs, texts, critical articles, bibliography, biography.

Also contains a wide selection of quotations. "Highly recommended as both an introduction and a tribute to this remarkable writer."—*Washington Post* "The most detailed, probing and comprehensive study of Ballard on the market."—*Boston Phoenix*.
8½x11", 176 pp, illus. PB. **$20 (last copies; order soon!)**

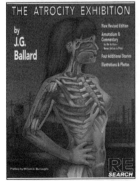

Atrocity Exhibition A dangerous imaginary work; as William Burroughs put it, "This book stirs sexual depths untouched by the hardest-core illustrated porn." Amazingly perverse medical illustrations by Phoebe Gloeckner, and haunting "Ruins of the Space Age" photos by Ana Barrado. Our most beautiful book, now used in many "Futurology" college classes. 8½x11", 136 pp, illus. PB **$20.**
LIMITED EDITION OF SIGNED HARDBACKS $150 (Only 20 copies left)

SPECIAL OFFER
ALL FOUR BALLARD BOOKS PLUS SEARCH & DESTROY #10 TABLOID ISSUE (WITH BALLARD INTERVIEW) FOR JUST $65 PLUS SHIPPING $10 DOMESTIC USA, $30 OVERSEAS